Evolution
Without
Evidence

Evolution Without Evidence

Charles Darwin and *The Origin of Species*

Barry G. Gale

University of New Mexico Press
Albuquerque

Library of Congress Cataloging in Publication Data

Gale, Barry G. (Barry George), 1941–
    Evolution without evidence.

    Bibliography: p.
    Includes index.
    1. Darwin, Charles, 1809-1882. The origin of
species. 2. Darwin, Charles, 1809-1882. 3. Naturalists—
England—Biography. I. Title.
QH365.08G34   575'.0092'4 [B]   82-4778
ISBN 0-8263-0609-8   AACR2

Manufactured in the United States of America.
Library of Congress Catalog Card Number 82–4778.
International Standard Book Number 0–8263–0609–8.
First edition

To the memory of my mother,
To my father,
And to the joy that is Heather

Contents

Acknowledgments

Many people have helped me in the years it took me to put this work together. I want to thank them for their patience and their contributions. Of these, I think I owe special thanks, and special gratitude, to Dr. Susan F. Cannon, formerly Curator of the History of Classical Physics and Geosciences at the Smithsonian Institution in Washington, D.C. She asked the toughest questions. Her recent untimely death has saddened the world of Darwin scholarship.

Evolution
Without
Evidence

Idling reader, you may believe me when I tell you that I should have liked this book, which is the child of my brain, to be the fairest, the sprightliest, and the cleverest that could be imagined, but I have not been able to contravene the law of nature which would have it that like begets like.

Cervantes, *Don Quixote*

1

A Theory by Which to Work

In Darwin's *Autobiography*, written between 1876 and 1882 for his family to read, he relates how, following his return to England from the *Beagle* voyage (1831–36), he set out "on true Baconian principles" and "without any theory" to collect facts "on a wholesale scale" regarding "the variation of animals and plants under domestication and nature." "I soon perceived that selection was the keystone of man's success in making useful races of animals and plants. But how selection could be applied to organisms living in a state of nature remained for some time a mystery to me."[1]

On September 28, 1838, the answer came to him from an unexpected source. As Darwin later recalled: "I happened to read . . . Malthus on *Population*, and being well prepared to appreciate the stuggle for existence which everywhere goes on from long-continued observation of the habits of animals and plants, it at once struck me that under these circumstances favourable variations would tend to be preserved, and unfavourable ones to be destroyed. The result of this would be the formation of new species. Here, then, I had at last got a theory by which to work; but I was so anxious to avoid prejudice, that I determined not for some time to write even the briefest sketch of it."[2]

Considerable argument exists among Darwin scholars regarding the exact role that Malthus played in Darwin's initial formulation of his natural selection theory. The old view takes Darwin at his word: Malthus's influence is seen to be crucial. Malthus enabled Darwin to see the force by which evolution could occur. Darwin had little if any conception of such a force before 1838.

3

The more recent view is that Malthus's influence has been over-rated—at best, Malthus served merely as a catalyst in the de-velopment of Darwin's ideas; at worst, Malthus played only a confirmatory role, for Darwin had his theory pretty much intact before he ever read Malthus's essay.[3]

I have written elsewhere on this issue, basing my views on an analysis of the evolving nature of Malthus's work and Darwin's "selective" reading of the sixth edition.[4] I tend to think that there is still much merit in the old view of Malthus's influence. On the whole, however, I think the subject has been overworked. I would like to change focus a bit: away from the inspiration for Darwin's theory to the theory itself. My initial questions are: What exactly did Darwin possess, in terms of a theory of evolution, by the end of 1838? How did he view that theory? And what did he do with it?

Many Darwin scholars analyzing this period seem to imply that October 1838 marks an endpoint in the development of Darwin's ideas—that by that date his theory was pretty much complete. Although several important theoretical breakthroughs were to be made later—most notably Darwin's concept of divergence of char-acter—the next twenty years are seen as a period during which Darwin's ideas were expanded and only certain minor odds and ends were tidied up.[5] After reviewing the events of 1838, Darwin biographers usually deal with Down House and Marriage, talk of Darwin's illness, touch briefly on the 1842 and 1844 *Essays*, de-scribe the rigid living and working schedule at Down, and pick up Darwin's evolutionary work in detail again when the Wallace Thunderbolt strikes. The more biologically knowledgeable of these biographers include a section analyzing Darwin's work on Cirri-pedes.[6]

Viewed in this way, the period often seems a time of puzzling delay. If Darwin had his theory pretty much intact by 1838, wrote it out in 1842, and expanded it in 1844, why did he take so long to publish?[7] The more psychologically oriented of Darwin com-mentators are inclined to scrutinize Darwin's psyche and look with suspicion at his relationship with his father.[8]

Viewing 1838 as an endpoint (or as more of an end than a begin-ning) is valuable in the sense that it helps focus our attention on Darwin's early, highly creative period of discovery following his return from the *Beagle* voyage. Yet, focusing exclusively this way tends to obscure the equally important and creative period that followed his reading of Malthus. It obscures the quite enormous effort that still needed to be undertaken before Darwin's ideas would be in a form acceptable to himself, or in a form having any chance of being acceptable to others. And it obscures what I have come to see as the crucially important organizational and management aspects of Darwin's achievement.

I want to look at Darwin's theory during this time (the 1838 period and beyond) as if the *Origin* did *not* follow—or, better still, with no idea at all as to what was to follow. Once we do this, then I think a whole series of different kinds of questions become important. We no longer ask: How did Darwin develop his theory? What role did Malthus play in its development? We now ask: How did Darwin view this theory? Was it important to him at the time? Was he convinced that it was true? Did he think he could prove it to other people? What sort of skills would be needed in order to do so? What was the state of Darwin's knowledge at this time? What relative importance did he assign to his evolutionary work vis-à-vis his work in other natural history areas—for example, geology? What was Darwin planning to do next?

Exploring questions such as these will lead, as I hope to demonstrate in the course of this book, to several conclusions quite different from those now generally held regarding Darwin and his work.

We will see, in terms of the development of his theory, how 1838 marked not an end but an important beginning for Darwin; how Darwin was ill-equipped in terms of knowledge and experience for the tasks that lay ahead of him at the time; the fact that for many years his species work was only a side interest and that he returned full-time to biological subjects (and indirectly to his species work) only after his career in geology seemed at a standstill; how really dependent he was on the help of others for much of his work on species, how well-developed organizational, management, and interpersonal skills and abilities helped him first secure and then benefit from this help. We will also see his doubt and uncertainty about his theory, owing, in part, to his failure to win important converts to his ideas even as late as 1859 (such close friends as the geologist Charles Lyell, the American botanist Asa Gray, and the biologist Thomas Henry Huxley remained unconvinced); and, finally, we will see Darwin's additional con-

cern about the weakness of the argument of the *Origin;* the fact
that in the *Origin* Darwin was forced to defend not so much the
correct theory as the least objectionable one; and that, far from
delaying publication of his ideas as some scholars have suggested,[9] given the quality of his evidence and the nature of his theory, Darwin was probably forced to publish his ideas too soon.

2

Before Malthus

Studying Natural History

Avoiding Prejudice

After reading Malthus, as we have seen, Darwin "had at last got a theory by which to work." But he was "so anxious to avoid prejudice" that he "determined not to write for some time even the briefest sketch of it." Let us look at these two passages in more detail.

When Darwin says that after reading Malthus he had "at last got a theory by which to work," I believe he means that he had at last got a working hypothesis—a theory that made some sense to him at the time but also a theory that needed to be tested and substantiated. His ideas at this time were nowhere near final formulation.

The second passage is less clear. So anxious was he "to avoid prejudice" that he "determined not to write for some time even the briefest sketch of it." But why would he need to avoid prejudice, and what sort of prejudice? And, even if he wanted to avoid prejudice, why would that desire lead to his determination "not to write for some time even the briefest sketch of it?" He could certainly have written a brief sketch without showing it to anyone.

Howard Gruber has written on the role that the threat of persecution might have played in Darwin's thinking. Gruber implies that Darwin's fear of persecution probably inhibited the free communication of his evolutionary ideas—contributing to his "long delay" in publishing his ideas—and that that fear was based in reality.[1]

Prejudice on the subject of evolution was undoubtedly plentiful at the time as well as later (as Darwin was to find out to his dismay following publication of the *Origin*), and Darwin would have had to be dull of mind to have missed it. I think, however, that negative reaction to unpopular, materialist views is only part of what Darwin was anxious to avoid. Another important potential source of his anxiety—and one also based very much in reality—can be found in Darwin's fear regarding his lack of natural history skills, experience, and competency at this time, and the resulting questionable quality of his theory.

The problem confronting Darwin at the end of 1838 was not so much the fact that if he communicated his ideas he would be severely criticized, but rather the fact that he did not have very much to communicate. His theory had, in essence, preceded his knowledge—that is, he had hit upon a novel and evocative theory of evolution with limited knowledge at hand to satisfy either himself or others that the theory was true. He could neither accept it himself nor prove it to others. He simply did not know enough concerning the several natural history fields upon which his theory would have to be based.[2]

Also, I think that from Darwin's perspective at the time, his theory *qua* theory was rudimentary and weak. It was not something he would have liked to defend, even if he had to. It was the sort of theory one puts in a desk drawer—to be looked at much later, to be written later still. And if it were the sort of theory that might be looked at unfairly at first—not given a fair chance because of the prejudices it might arouse—all the more reason to take care with it, not, so to speak, to let it out of the bag too soon. It was in this sense, it seems, that it was a theory he "determined not even to write the briefest sketch of." It was a very uncertain theory on a highly controversial subject.[3]

Soon after its initial formulation, Darwin's theory and his work on species took on a very distant aspect. To his cousin and fellow naturalist, W. Darwin Fox, in January 1841, Darwin described his activities regarding species and varieties as a "some-day work . . . the smallest contributions thankfully accepted";[4] on October 12, 1845, Darwin wrote to his zoologist friend Leonard Jenyns that his work on species was in process, but "I shall not publish on this subject for several years";[5] and again to Jenyns about the same year, Darwin talked of his "far distant work on species."[6]

These statements do not indicate a man very certain of his work at the time.

The fact is that in 1838 Darwin's knowledge and experience in natural history were decidedly limited, and this was especially true with regard to zoology, botany, geographical distribution, and paleontology, several of the more important fields from which evidence in support of his theory would have to be gathered. The idea that Darwin returned from the *Beagle* voyage a finished naturalist ready to take the British natural history community by storm is simply not true, though he was certainly by that time a promising young naturalist.[7] His development, though steady, was slow and meandering. Except for his five years on the *Beagle*—when he was forced to attend seriously to several natural history areas—he had not systematically studied natural history subjects.

But let us look in more detail at his natural history education prior to 1838.

Dr. Butler's and Edinburgh

Darwin seems to have been interested in the study of nature, in one form or another, since his early youth. By the time he entered Shrewsbury Day School in 1817 his taste for natural history, and especially collecting, was already well developed. He collected shells, seals, franks, coins, and minerals. He also tried to make out the names of plants. When he was sent a year later (1818) to board at Dr. Butler's "great school in Shrewsbury," his natural history interests had not abated: he collected minerals; he collected and observed insects (only dead ones, on the advice of his compassionate sister); and he read Gilbert White's *Natural History of Selborne* and wondered why "every gentleman did not become an ornithologist."[8] These interests were outside the normal course of study at Dr. Butler's, which was "strictly classical," except for some ancient geography.

Darwin read a lot during this period—Shakespeare, Byron, Scott —and he developed a passion for shooting, which would stay with him through all the years of his formal schooling and some years beyond. "I do not believe that anyone could have shown more zeal for the most holy cause than I did for shooting birds," he later recalled.[9]

Whether shooting or natural history interests (or reading, for that matter) were more important to him at this time is difficult to say. One thing is certain: Darwin had an active dislike for the formal academic course at Dr. Butler's and sought an outlet for his passions and intellectual interests elsewhere. Natural history subjects became one of these extracurricular areas of interest.

As Darwin was not doing well academically at Dr. Butler's, Robert Darwin, Charles's father, sent his son in 1825, at the early age of sixteen, to study medicine at Edinburgh University. His brother Erasmus was already there. The Edinburgh medical course was different from Dr. Butler's "strictly classical" curriculum in that at Edinburgh natural history subjects were part of the required course of study. According to Darwin's class cards, in his first year he had Thomas Hope for chemistry, Andrew Duncan for materia medica, Alexander Monro for two courses—one on anatomy, physiology, and pathology, the other on the principles and practice of surgery—and Drs. Graham and Alison for clinical lectures. In his second year (his class cards for this year are not preserved, so less is known) he enrolled in courses on the practice of physic and midwifery, a class on natural history, and Robert Jameson's course on geology and zoology.[10]

Of all the lecture courses, he found only Hope's on chemistry interesting; the others were "intolerably dull" and Jameson's on geology and zoology he found so boring that he determined "never as long as I lived to read a book on Geology or in any other way to study the science."[11]

Thus even at Edinburgh, where several natural history subjects were part of the curriculum, he was not interested in the academic course. I suspect that the specific subject matter had little to do with Darwin's dislike for it; the fact that it was an academic lecture course itself seems to have been enough to turn him sour. Perhaps Darwin did not like being lectured to.[12]

As he had at Dr. Butler's, at Edinburgh Darwin sought fulfillment of his natural history interests outside the formal academic environment. Darwin became a member of the Plinian Society, before which he presented two short biological papers. The society seemed to provide a vehicle for the expression of his natural history zeal, and he took an active interest in its activities. The society minutes reveal that a week after Darwin was elected a member, he was chosen as one of the five to sit on the Plinian ruling council; they also indicate that while he was at Edinburgh he was present at eighteen of the nineteen society meetings held during that time. In addition, Darwin is listed as having participated in four of the meeting discussions—one on the subject of natural classification and specific characters—though there is no mention of what he said.[13]

Also during this time, both within and outside the Plinian Society, Darwin became acquainted with a group of young men (most were his senior in age) who shared his interests in natural history. There were William Ainsworth (geology), Dr. John Coldstream (zoology), George Fife (medicine), William Kay (medicine), William Macgillivray (zoology and assistant keeper of the Museum of Natural History), and Dr. Robert Grant (zoology). Darwin often accompanied Grant to tidal pools to search for marine specimens. On one walk with Grant, Darwin heard him "burst forth in high admiration of Lamarck and his views on evolution." Darwin "listened in silent astonishment" but without any effect on his mind. He had previously read his grandfather Erasmus Darwin's *Zoonomia*, where similar views were upheld, but also without any lasting effect.[14]

To further his natural history interests at Edinburgh, Darwin joined the Royal Medical Society and attended, with Grant, various Wernerian Society meetings, where papers on natural history were read, discussed, and later published in the society's *Transactions*. Darwin also attempted to learn taxidermy and seems to have had considerable contact with William Macgillivray, probably on the subject of ornithology, given Macgillivray's primary interests.[15]

The only indication we have of any detailed natural history research during Darwin's Edinburgh days (in addition to his two small papers delivered before the Plinian) is a notebook he kept between March 16 and April 23, 1827, in which he recorded various observations and described dissections of various marine animals.[16]

For example, on March 16, 1827, the notebook indicates that Darwin procured from the black rocks at Leith a large lumpfish which he and Dr. Grant dissected; on the same date he caught a small, green Aeolis and a Tritonia and examined the ova of the Purpura Lapillus, appending a sketch; two days later (March 18) he sketched three organisms "found growing out of Alcyonium";

on March 19 he observed (among other things) the ova in the Flustra Foliacea and Truncata; on April 15 he noted and sketched "whitish circular masses of ova" on Fucus and also one of the ciliated embryos, and so on.[17]

Although it is clear that Darwin had at least some experience with the description and dissection of organisms early in his education, he later felt that his experience in this area had been inadequate. In his *Autobiography* he recalled with disappointment his early efforts at dissection and drawing: "It has proved one of the greatest evils in my life that I was not urged to practice dissection, for I should have got over my disgust; and the practice would have been invaluable for all my future work. This has been an irremedial evil as well as my incapacity to draw . . . from not having had any regular practice in dissection, and from possessing only a wretched microscope my attempts [during this time] were very poor."[18]

Although Darwin did not seem particularly interested in the study of medicine, his two years at Edinburgh did provide him with a broad exposure to several natural history subjects—geology (in a negative way), zoology (marine biology, ornithology), taxidermy, evolutionary ideas (through Grant), and presumably many other subjects discussed at the Plinian, Royal Medical, and Wernerian Society meetings. As yet he showed no preference for any one particular area of study.

Also, at Edinburgh, Darwin first encountered some unpleasant experiences associated with his interests in natural history. In addition to his dislike of geology, which might be attributed more to his dislike of Jameson's lectures than to any antipathy to the subject, and his realization that he was not very competent at two important skills useful for work in natural history, dissecting and drawing, he had a disagreeable experience with Grant on a matter of scientific priority. One of Darwin's discoveries regarding the Flustra larvae so excited him that he rushed to tell Grant of his finding. Grant rebuked him for working on a subject that he was also studying at the time, and he asked Darwin not to publish his results. The sharp edges of this incident were later smoothed somewhat when Grant mentioned Darwin's work in the area in his memoir on Flustra.[19]

Although Darwin during these years showed great interest, and at times some zeal, in natural history subjects, an equally impor-

tant interest lay quite outside the realm of science—his passion
for shooting. As he later remembered: "The autumns were devoted to shooting. . . . My zeal was so great that I used to place
my shooting boots open by my bed-side when I went to bed, so
as not to lose half-a-minute in putting them on in the morning;
and on one occasion I reached a distant part of the Maer estate
[his uncle Josiah Wedgwood's home, where he often went to shoot]
on the 20th of August for black-game shooting, before I could see."
His passion was so great that he began to feel guilty. "How I did
enjoy shooting, but I think that I must have been half-consciously
ashamed of my zeal, for I tried to persuade myself that shooting
was almost an intellectual amusement; it required so much skill
to judge where to find most game and to hunt the dogs well."[20]

Cambridge

Darwin's aversion to medicine and his poor record at Edinburgh
convinced his father to remove him after his second year and send
him to Cambridge, to prepare for a career in the clergy. Robert
Darwin seemed determined to find some proper profession for
his son.

Darwin entered Cambridge in 1828 and had, by his own admission, a very happy, if not productive, time. There he led the life
of the well-to-do English country gentleman. He joined a "sport-
ing set" that did a good deal of drinking, hunting, and riding
cross-country. He became involved in a musical group and sometimes hired the chorister boys from King's College Chapel to sing
in his rooms. He became interested in engravings, purchased several drawings, and frequented the Fitzwilliam Museum. He became a member of a Gourmet Club, which met in weekly feasts.
He attended many suppers, parties, and gala events. And he studied when he had to—in order to prepare for exams.[21]

His college was Christ's, which had the reputation then of being
a "pleasant, fairly quiet college, with some tendency towards
'horsiness'; many of the men made a custom of going to New-
market during the races." It was a not "unpleasant college for
men with money to spend and with no great love of strict disci-
pline."[22] Darwin fit in well.

With regard to course work, he attempted mathematics, but

without much success. He did not attend Professor Adam Sedgwick's "eloquent and interesting" lectures on geology, so bored had he been by Jameson's lectures at Edinburgh. He later regretted this, saying that had he attended "I should probably have become a geologist earlier than I did."[23] He did attend Professor John Stevens Henslow's lectures on botany, which he liked, but he "did not study botany."[24] However, he did accompany Henslow's course on several of its field trips and found great pleasure in these. Upon graduation, he finished tenth on the list of *hoi polloi*, or men who do not go in for honors—not a poor showing though certainly not spectacular.

His most avid natural history pursuit at Cambridge was collecting beetles, which he did with great zeal. This is reflected in contemporary letters to his cousin and Christ's colleague, W. Darwin Fox. In anticipation of a new college term, Darwin wrote to Fox in January 1829: "How we would talk, walk and entomologise!";[25] and a month later: "The first two days I spent entirely with Mr. Hope and did little else but talk about and look at insects."[26] In the summer of 1829, Darwin was going to Maer, he informed Fox, "in order to entomologise."[27] In March of 1830, Darwin wrote concerning Fox's imminent arrival in Cambridge: "What fun we will have together; what beetles we will catch."[28] And, in November of 1830, he lamented: "I have not stuck an insect this term, and scarcely opened a case."[29]

Darwin's passion for collecting beetles was a passion for collecting, for he "did not dissect them and rarely compared their external characteristics with published descriptions, but got them named anyhow."[30] His bad experience with dissecting at Edinburgh may have made him shy away from it at Cambridge.

As he had at Edinburgh, at Cambridge Darwin developed a number of close friendships with others in natural history. The most important of these was with Henslow. He later felt that his friendship with Henslow "influenced my whole career more than any other" circumstance.[31] It was Henslow, as we shall see, who recommended that Darwin begin the study of geology and arranged for him to accompany Professor Sedgwick on a geological tour of North Wales. It was Henslow who was responsible for Darwin receiving the offer of the position of naturalist aboard the *Beagle*. It was Henslow who provided Darwin with guidance and information throughout the voyage and became the recipient of Darwin's specimens sent from abroad. After the voyage, it was Henslow who helped Darwin secure money from the British Treasury for the volume on the zoology of the *Beagle* voyage. And it was Henslow who read the proofs of Darwin's *Journal of Researches* prior to publication, recommended Darwin for membership in the Geological Society, and was a constant source of information to Darwin on questions related to botany and other natural history subjects.

Henslow, whose knowledge, Darwin tells us, "was great in botany, entomology, chemistry, mineralogy, and geology,"[32] was a highly respected professor of mineralogy and later botany at Cambridge, with a talent for lecturing and an enthusiasm for the study of natural history which was infectious. He had a gift for drawing young students interested in natural history around him. At the time Darwin entered Cambridge, Henslow was at the height of his powers; his home served as one of the centers of natural history and philosophy interests and activities, both for students and faculty. "He kept open house every week, where all undergraduates and several older members of the University, who were attached to science, used to meet in the evening."[33] Through Fox, Darwin was able to receive an invitation to these affairs (they seemed to be as social as they were scientific), and before long he became "well acquainted" with Henslow. During the latter half of his stay at Cambridge, Darwin became very close to Henslow, taking long walks with him on most days. He was called by some of the dons "the man who walks with Henslow."[34]

In Darwin's last year at Cambridge, he read Humboldt's *Personal Narrative* and Sir John Herschel's *Introduction to the Study of Natural Philosophy*. These studies taken together, he later remembered, stirred in him "a burning zeal to add even the most humble contribution to the noble structure of Natural Science."[35]

In the summer of 1831, Henslow persuaded Darwin "to begin the study of geology" and arranged for Darwin to accompany Sedgwick on a geological tour of North Wales. This tour was significant as a training ground for the geological excursions Darwin would undertake during the *Beagle* voyage. Sedgwick was then one of the most prominent geologists in his field. He had just finished two terms as president of the Geological Society of London. Darwin could not have had a better geological beginning.

Paul Barrett has written of this excursion and has reproduced for the first time Darwin's geological notes of the trip and two letters from Sedgwick to Darwin connected with the expedition. Barrett feels that these notes will help provide "insight into both Darwin's geological knowledge and scientific philosophy at the time." Barrett sees them as indicative of Darwin's "thorough knowledge of geology" and "geological competence at this time." He goes so far as to contend that because of the quality of the notes "we may . . . assume . . . that, although repulsed by Jameson, he [Darwin] had at Edinburgh nevertheless learned well the fundamentals of both descriptive and theoretical geology."[36]

The notes, although highly descriptive and seemingly thorough, do not appear to represent more than a rudimentary understanding of geology, though I make no claim to any great knowledge of geology. What disturbs me about Barrett's claim is that he fails to consider other evidence which would suggest that though Darwin might have had some knowledge of geology at the time, he was really at the beginning of any serious study of the subject. For example, in his *Autobiography* Darwin comments that at this time (following graduation from Cambridge in April 1831) Henslow "persuaded me *to begin* the study of geology."[37] We have already seen Darwin lament about not having attended Sedgwick's lectures, for if he had "I should probably have become a geologist earlier than I did." I interpret this to mean that while at Cambridge Darwin was *not* a geologist. Barrett's statement about Darwin learning "well the fundamentals of both descriptive and theoretical geology" while at Edinburgh does not make much sense in light of Darwin's lament. Also, in the *Autobiography* Darwin refers to the excursion with Sedgwick as being of "decided use in teaching me *a little how* to make out the geology of a country."[38] This seems to suggest Darwin was just beginning to learn how to do that.

Finally, there is evidence that in July 1831 (the expedition to North Wales left Shrewsbury in August 1831) Darwin, following Henslow's suggestion, was working hard to learn geology, especially those aspects of it related to field work. We find that at the time Darwin was "working like a tiger" at geology but not finding it "as easy as I expected."[39] Writing to Henslow on July 11, 1831, Darwin provided the following account of his activities: "I should have written to you some time ago, only I was determined to wait for the Clinometer, & I am very glad to say I think it will answer admirably: I put all the tables in my bedroom, at every

conceivable angle & direction. I will venture to say I have measured them as accurately as any Geologist going could do. . . . I
have been working at so many things: that I have not got on much
with geology: I suspect the first expedition I take, clinometer &
hammer in hand, will send me back very little wiser & [a] good
deal more puzzled than when I started.—As yet I have only in-
dulged in hypotheses. . . . "[40] This sounds to me like the report
of someone in the process of learning geology.

Following Darwin's return from North Wales at the end of
August, he entered into his "other" life—shooting (though it
might be more accurate to say that natural history was his "other"
life at this time). As Darwin later recalled, putting things, perhaps, in proper historical perspective, following the Sedgwick expedition he returned to Shrewsbury and to Maer for shooting; "for
at that time I should have thought myself mad to give up the
first days of partridge-shooting *for geology or any other science.*"[41]

To summarize, Darwin's natural history activities at Cambridge
gave him at best a broad but rudimentary acquaintance with geology and botany, perhaps a more intimate appreciation of entomology (collecting rather than analysis), and an enthusiasm for, but not
apparently a well-founded understanding of, a broad range of natural history subjects. His identity at Cambridge was undoubtedly
that of an entomologist of the collector sort. It was only after
graduation from Cambridge, in the summer of 1831, that Darwin's
interests began to turn slowly to geology.

The Voyage of the Beagle

In August 1831, Henslow received a letter from George Peacock,
Lowndean Professor of Astronomy at Cambridge, regarding an expedition of *H.M.S. Beagle* to South America, the South Sea Islands,
and around the world via the Indian Archipelago. The expedition
was to be headed by a Captain Robert Fitzroy. The primary purpose of the voyage was to survey unchartered coastal areas in South
America. ". . . The vessel is fitted out expressly for scientific purposes, combined with the surveys: it will furnish therefore a rare
opportunity for a naturalist & it would be a great misfortune that
it should be lost," Peacock wrote. "An offer has been made to
me," he continued, "to recommend a proper person to go out as

a naturalist with this expedition: he will be treated with every consideration: the Captain [i.e., FitzRoy] is a young man of very pleasing manners (a nephew of the Duke of Grafton) of great zeal in his profession & who is very highly spoken of: if Leonard Jenyns could go, what treasures he might bring home with him . . . in the absence of so accomplished a naturalist, is there any person whom you could strongly recommend: he must be such a person as would do credit to our recommendation." "Do think on this subject," Peacock added, "it would be a serious loss to the cause of natural science, if this fine opportunity was lost."[42]

Leonard Jenyns, at first inclined to go, decided against it. Then Henslow himself thought of going, but eventually backed down. On August 24, 1831, Henslow wrote to Darwin offering him the *Beagle* position. This letter is significant in that, after providing Darwin with an outline of the voyage's purpose and direction, Henslow provided a candid estimate of Darwin's natural history capabilities at the time: "I have stated that I consider you to be the best qualified person I know of who is likely to undertake such a situation. I state this not on the supposition of your being a *finished* Naturalist, but as amply qualified for collecting, observing, and noting anything new to be noted in Natural History. . . . Don't put on any modest doubts or fears about your disqualifications for I assure you I think you are the very same man they are in search of. . . ." Henslow chose his words carefully, for it was only in the sense that Jenyns and he had already refused the position that Darwin was the best qualified person likely to accept. "Capt. F." Henslow wrote reassuringly at another point, "wants a man (I understand) more as a companion than a mere collector & would not take anyone however good a Naturalist who was not recommended to him likewise as a *gentleman*."[43] Darwin certainly met the latter qualification, even though he might be somewhat shaky on the former.

What followed is most familiar by now. Robert Darwin, concerned for the future of his son and feeling that, among other dark prophecies, the voyage was, as he termed it, a "wild scheme" and would be disreputable to Charles's character as a clergyman thereafter, at first refused permission for his son to go. Soon, however, he relented under pressure from Darwin's uncle, Josiah Wedgwood. On December 27, 1831, the *Beagle* set sail from Plymouth harbor.

In looking back on the events of his life, Darwin saw the voyage of the *Beagle* as the most important: "The voyage of the *Beagle* has . . . determined my whole career."[44] Yet at the beginning Darwin's role on the voyage was not very clear.

Jacob Gruber has argued that the *Beagle* already had a naturalist, the ship's surgeon, Robert McCormick, and that Darwin was there really as a private person. The role of surgeon as naturalist in the expedition, Gruber tells us, was part of a developing tradition in government-sponsored scientific research. McCormick was such a surgeon/naturalist. McCormick, however, soon left the expedition under strained circumstances, perhaps in part because of Darwin's presence, though this is not clear. The fact that Darwin had control over his own collections, paid his own expenses, had his own servant, and could leave the voyage any time he wanted, further suggests to Gruber the "private" nature of his activities.[45]

However "private" or "official" the position might have been, FitzRoy (from what can be determined from Henslow's letter) seemed more concerned about who the person was, than about the auspices under which he came, or about what the individual would do. FitzRoy seemed to want a gentleman-companion— someone he could talk to. The *Beagle*'s officers could certainly fit that bill, although, because of his position, FitzRoy's relationship to them would of necessity be more formal. We know that Darwin almost lost the position when it looked as if one of FitzRoy's close, nonnaturalist friends might accompany him.[46] This gives us some indication of the sort of person FitzRoy was looking for. The fact that the naturalist would alone share the Captain's mess further suggests the prerequisite of social and personal acceptability as qualification for the position. From FitzRoy's perspective, then, the naturalist on the *Beagle* would be a companion first, and a scientist second.

From Peacock's perspective (and probably Henslow's too) Darwin's role in the *Beagle* was to bring back natural history "treasures" to England—to be, in essence, a collector of specimens for study by naturalists back home. As Peacock pointed out to Henslow in a follow-up letter: "What a glorious opportunity this would be for forming collections for our museums."[47] Other British scientists seemed to have shared Peacock's view of the opportunities the *Beagle* voyage afforded for new collections of specimens;

Darwin was well supplied with equipment, instructions, and advice from a good number of all-too-eager-to-help naturalists before he departed. "I feel my blood run cold at the quantity I have to do," Darwin wrote his sister Susan on September 9, 1831, but ". . . everybody seems ready to assist me. The Zoological [Society] wants to make me a corresponding member."[48] And a month later he asked FitzRoy: "Have you a good set of mountain barometers? Several great guns in the scientific world have told me some points in geology to ascertain which entirely depend on their relative height."[49] By the time he had departed England, he could write confidently to his father that "no person ever went out better provided for collecting and observing in the different branches of Natural History. In a multitude of counsellors I certainly found good."[50]

From Darwin's perspective, the voyage seemed a wonderful opportunity to travel to exotic places and enjoy the luxuriance of tropical vegetation. There would certainly be many interesting things to do: his natural history work; learning navigation and meteorology; conversing with a "pleasant set of officers," which he was sure there would be. He could even hope for new and exciting shooting adventures. After meeting with FitzRoy and being given assurances as to the position, Darwin wrote to Henslow on September 5, 1831: "till one today I was building castles in the air about hunting Foxes in Shropshire, now Llamas in S. America."[51]

The role that Darwin did in fact play as a naturalist aboard the *Beagle* was closest to the one originally envisioned by Peacock. Darwin barely got along with FitzRoy, who, though usually considerate of him, would at times go into fits of deep melancholy and be almost impossible to deal with. Darwin became primarily a scientist and secondarily a companion to FitzRoy. Darwin did get to see exotic places and tropical vegetation (as his *Journal of Researches* attests)[52] but he also had to work harder than he had ever before in his life. He soon gave up shooting. He delegated that responsibility to a servant, Sym Covington, whom he hired during the course of the voyage. I do not think he ever seriously attempted to learn navigation and meteorology, but I may be wrong. He did get along well with the *Beagle* officers and crew, although feverish work on his collections or seasickness left him little time to socialize.

In the middle of the voyage, when Darwin was beginning to

receive some recognition back in England for his work, he wrote
to a Cambridge friend, J. M. Herbert, concerning his role on the
voyage. He spoke very frankly: "You rank my Natural History
labours too high. I am nothing more than a lion's provider: I do
not feel at all sure that they will not growl and finally destroy
me."[53] The lions were the naturalists back home. Darwin was
their "collector." They had the knowledge and he collected specimens for them to analyze. It was a rather finely circumscribed
role which particularly suited his limited skills and knowledge
at the time.

Henslow was the lions' representative. He helped direct Darwin's
collecting activities. This involved doing two things—comment-
ing on the value of what Darwin had collected and providing Darwin with guidance for further collecting. Henslow also played the
role of Darwin's agent in England, serving as the recipient and
caretaker for all specimens sent from abroad, and also providing
Darwin with books, news, and other things when requested.

In a series of letters exchanged between the two men during
the *Beagle* years, we get a sense of the relationship that developed. In a letter dated October 18, 1831, Darwin asked Henslow
to serve as his agent for the receipt of shipments of specimens
from abroad: "I seize the opportunity of writing to you on the
subject of consignment.—I have talked to everybody: & you are
my only recourse. . . ."[54] Henslow agreed. On November 20,
Henslow provided Darwin with proofs of a recently written paper
which he thought might "be of service in directing your attention whilst collecting land shells."[55] On May 18, 1832, Darwin
informed Henslow of a delay in the first shipment of specimens
to England and expressed concern over whether he had done a
competent collecting job: "I have determined not to send a box
till we arrive at Monte Video . . . —I am afraid when I do send it,
you will be disappointed, not having skins of birds & but very
few plants, & geological specimens small. . . ."[56] In August, Darwin was still anxious: "And now for an apologetical prose about
my collection,—I am afraid you will say it is very small,—but I
have not been idle . . . The box contains a good many geological
specimens . . . I have come to the conclusion that 2 animals with
their original colour & shape noted down, will be more valuable
to Naturalists than 6 with only dates and place—I hope you will
send me your criticisms. . . ."[57]

In November 1832, Darwin, showing considerable zeal for col-

lecting, alerted Henslow to French attempts to get good specimens: "By ill luck the French government has sent one of its
Collectors to the Rio Negro—where he has been working for the
last six months. . . . So that I am very selfishly afraid that he will
get the cream of all the good things before me."[58]

In January 1833, Henslow, having received the first shipment
of specimens from England, offered Darwin guidance: "I would
not bother myself about whether I were right or wrong . . . note
all that *may* be useful. . . . So far from being disappointed with
the Box—I think you have done wonders." He then added specific
instructions for specimens that should be collected and how these
should be packed prior to shipment: "Avoid sending *scraps.* Make
the specimens as perfect as you can, *root, flowers* & *leaves* &
you can't do wrong. In large ferns & leaves fold them back upon
themselves on *one* side of the specimen & they will fit into a
proper sized paper. . . . L. Jenyns does not know what to make of
your land Planariae. Do you mistake for such the curious Genus,
'Oncidium' allied to slug, of which fig. is given in Linn. Trans-
act. & one not the marine species also *mullusca,* perhaps Doris
& other genera—Specimens & observations upon these wd. be
highly interesting. . . . And now for the Box—Lowe *under packs*
Darwin *overpacks*—The latter is in fault on the right side. You
need not make quite so great a parade of tow & paper for the geolog.
specimens, as they travel very well provided they be each wrapped
up *german fashion* & closely stowed—but *above all things* don't
put tow around *anything* before you have first wrapped it up in a
piece of thin paper. . . . ," and so on.[59]

In April, Darwin was asking for "any books, scientific travels
etc etc which would be useful to me."[60] And in August, Henslow
sent more instructions to help guide Darwin's collecting activities: "Send home every scrap of Megatherium skull you can set
your eyes on—& *all* fossils. Use your sweeping net for I forsee
that your minute insects will nearly all turn out new."[61] Henslow
also sent Darwin several books which he thought might be of
use.[62]

In March 1834, a happy Darwin, expressing delight that so much
interest had been shown back home in his specimens, declared
to Henslow: "I am indeed bound, & will pledge myself to collect
whenever we are in parts not often visited by Ships & Collectors. . . . "[63] In July, Henslow wrote with still more instructions:

". . . pray don't entirely neglect to dry plants—Those sent are *all* of the greatest interest—Send minute things . . . & common weeds & grasses, not to the neglect of flowering shrubs."[64] In the same month from Valparaiso, an excited Darwin wrote of a possible fossil cache: "I have just got scent of some fossil bones of a MAMMOTH! what they may be, I do not know, but," he added with pride and determination, "if gold or galloping will get them, they shall be mine."[65]

Finally, from Sydney in January 1836, Darwin described to Henslow his last large collecting effort—at the Galapagos Islands: ". . . I collected every plant, which I could see in flower. . . . I paid also much attention to the Birds, which I suspect are very curious."[66]

Later, in his *Journal of Researches*, published following the voyage, Darwin paid tribute to Henslow's assistance: ". . . My most sincere thanks to the Reverend Professor Henslow," Darwin wrote in his Preface, "who, when I was an under-graduate at Cambridge, was one chief means of giving me a taste for Natural History— who, during my absence, took charge of the collections I sent home, and by his correspondence directed my endeavours. . . . "[67]

Learning Geology

Although Darwin's primary role during the voyage was to collect specimens for others to analyze back home, he did develop, after embarking on the voyage, two special interests of his own. One was to study the geology of the various places visited and perhaps to write a book on the subject (this went far beyond the mere collection of mineralogical specimens); the other was to keep a detailed diary of the voyage, and perhaps to turn this into a published volume as well. As Darwin later recalled in his *Autobiography:* "It then [after surveying St. Jago in 1832, at the beginning of the *Beagle* voyage] first dawned on me that I might perhaps write a book on the geology of the various countries visited, and this made me thrill with delight. . . . Later in the voyage Fitz-Roy asked to read some of my Journal, and declared it would be worth publishing; so here was a second book in prospect!"[68] Darwin's desire to write a book on the geology of the countries visited by the *Beagle* was indicative of his new found interest in geology. This interest, during the course of the *Beagle*'s voyage, grew by leaps and bounds.[69]

In 1832, Darwin informed his father that he found "the Natural History of all these unfrequented spots [the Canary Islands and other places] most exceedingly interesting, especially the geology."[70] To Fox, in May of the same year, he was more enthusiastic: "My collections go on admirably in almost every branch . . . but Geology carries the day: it is like the pleasure of gambling. Speculating . . . what the rocks may be, I often mentally cry out 3 to 1 tertiary against primitive; but the latter have hitherto won all the bets."[71] In the same month, Darwin wrote to Henslow concerning St. Jago: "The geology was preeminently interesting, and I believe quite new: there are some facts on a large scale of upraised coast . . . that would interest Mr. Lyell." Later in the letter he added: "Tell Prof. Sedgwick he does not know how much I am indebted to him for the Welsh Expedition,—it has given me an interest in Geology—which I would not give up for any consideration. . . . "[72]

By midsummer 1834, Darwin was revealing, in a letter to his Cambridge friend C. Whitley, that he found geology "a never-failing interest . . . it creates the same grand ideas respecting this world which Astronomy does for the universe."[73] And a year later, he was proselytizing on its behalf: "I am glad to hear you have some thoughts of beginning Geology," he wrote to Fox. "I am become a zealous disciple of Mr. Lyell's views, as known in his admirable book. Geologising in South America, I am tempted to carry parts to a greater extent even than he does. Geology is a capital science to begin, as it requires nothing but a little reading, thinking, and hammering."[74]

In the summer of 1836, as the *Beagle* moved slowly toward England, Darwin was writing to Henslow from St. Helena asking to be proposed for membership in the Geological Society—"I am very anxious to belong. . . . "[75] By the time the *Beagle* docked in October 1836, the transformation of Darwin's natural history identity from entomologist to geologist was complete.

His attraction to geology was so strong during the voyage that it actually overwhelmed the greatest passion of Darwin's life to that point—his passion for shooting. That transformation marks, in my opinion, the change in Darwin from casual, gentlemanly observer of nature to serious student. "Looking backwards, I can now perceive how my love for science gradually preponderated over every other taste," Darwin later remembered in his *Auto-biography*. "During the first two years [of the voyage] my old passion for shooting survived in nearly full force and I shot myself all the birds and animals for my collection; but gradually I gave up my gun more and more, and finally altogether to my servant, as shooting interfered with my work, more especially with making out the geological structure of a country. I discovered, though unconsciously and insensibly, that the pleasure of observing and reasoning was a much higher one than that of skill and sport."[76]

It might be interesting to speculate as to why geology and not some other natural history field so attracted Darwin at this time. Certainly geology was then a controversial subject, probably much discussed. Also, Darwin had found in the first volume of Lyell's *Principles of Geology* a fascinating approach to the study of geology based on the "meaning" of the rocks. This seems to have been a very significant factor.

I think also Darwin's fascination for geology over other natural history subjects during the voyage was influenced by the various sorts of tasks he associated with each. For example, with regard to zoology and botany, Darwin's tasks were those of collection, dissection, and description. He found specimens (or had to shoot them), he described them (dissecting some of the marine invertebrates), and he packed them for shipment back home. There was little thinking and less speculation involved. He might even have disliked some of this work because of his lack of drawing and dissecting skills.

Geology set a different sort of—and I think a much more intellectually exciting and challenging—task for him. Although he collected minerals and fossils from the various geological formations he studied, and he described them in careful detail and prepared simple diagrams (he did rot have to dissect), this collecting and describing work was only a prelude to Darwin's major challenge with regard to geology—:o understand how things came to be as they were. It was the puzzle that each new formation presented and the need to try to solve that puzzle through a series of hypotheses that I think Darwin found so exciting. It was, as he wrote to Fox, "like the pleasure of gambling. Speculating . . . what the rocks may be." As Darwin later described the process of geological exploration in his *Autobiography*: "On first examining a new district nothing can appear more hopeless than the chaos of rocks; but by recording the stratification and nature of the rocks and fossils at any point, always reasoning and predicting what will be found elsewhere, light soon begins to dawn on the district, and the structure of the whole becomes more or less intelligible. I had brought with me the first volume of Lyell's *Principles of Geology*, which I studied attentively; and this book was of the highest service to me in many ways."[77]

Gaps in Knowledge

Although Darwin had found a fascinating field of study in geology and was beginning to take greater interest in other natural history areas as well, throughout the *Beagle* voyage his knowledge of these fields lagged far behind his enthusiasm for them. With regard to geology, for example, in the same letter to Fox in which Darwin declared his zealous support of Lyell's views, he also admitted: "I have a considerable body of notes together; but it is a constant subject of perplexity to me, whether they are of sufficient value for all the time I have spent about them. . . . " And to Henslow in March 1834, he lamented: "I have no books, which tell me much, and what they do I cannot apply to what I see. In consequence I draw my own conclusions, and most gloriously ridiculous ones they are. . . . Can you throw any light into my mind, by telling me what relation cleavage & planes of deposition bear to each other?"[78] And again to Henslow in July: "When I return to England, you will have some hard work in winnowing my Geology; what little I know, I have learnt in such a curious fashion, that I often feel very doubtful. . . . "[79] And to his sister Susan, on August 4, 1836, he wrote: "Professor Sedgwick mentioning my name at all gives me hopes that he will assist me with his advice, of which, in my geological questions, I stand in need."[80]

He was experiencing similar gaps of knowledge with regard to zoology. Concerning a bird he had shot, Darwin wrote to Henslow on November 24, 1832: "To my unornithological eyes, [it] appears to be a happy mixture of lark pidgeon & snipe . . . ; I suppose it will turn out to be some well-known bird, although it has quite baffled me. . . . "[81] And in a March 1834 letter to Henslow, in which he discussed his numbering of specimens, Darwin talked

of his "entire ignorance of comparative Anatomy."[82] This ignorance, combined with his poor dissecting skills, rendered questionable a good deal of the zoological work he had done during
the voyage. As he later remembered in his *Autobiography:* "From
not being able to draw and from not having sufficient anatomical knowledge a great pile of MS. which I made during the voyage has proved almost useless."[83]

Finally, with regard to botany, a letter to Henslow soon after
Darwin's return to England succinctly sums up the state of Darwin's botanical knowledge at the time: "You have made me known
amongst the botanists," he wrote on November 2, 1836, "but I
felt very foolish, when Mr. Don remarked on the beautiful appearance of some plant with an astoundingly long name, & asked me
about its habitation. Some-one else seemed quite surprised that I
knew nothing about a carex from I do not know where. I was at
last forced to plead most intire [*sic*] ignorance,—and that I knew
no more about the plants which I had collected, than the Man in
the Moon."[84]

The Post-Beagle Period

The *Beagle* voyage had given Darwin major new responsibili-
ties. He had worked hard to fulfill them. In the two years following his return to England, he worked even harder to sort, distribute,
and publish materials he had collected or worked on during the
voyage. It was one of the busiest, and most productive, periods of
his life.[85] His fascination with geology would continue and his
knowledge of this and other natural history fields would begin to
grow rapidly, though he would for some time still be dependent
on others (the "lions") for much of his knowledge in these areas.

The first order of business was to ensure that his specimens
were placed in competent hands. Because of limited knowledge,
he sought help with almost all the specimens he had collected,
including those in geology. "For even in Geology, I suspect much
assistance & communication will be necessary," he wrote to
Henslow.[86]

At first Darwin did not find many naturalists interested in his
collections, except the young Hunterian Professor at the Royal
College of Surgeons, Richard Owen (in anatomy), and Lyell and

the Geological Society's curator and librarian, William Lonsdale (in geology)—"If I was not much more inclined for geology, than the other branches of Natural History, I am sure Mr. Lyell's and Lonsdale's kindness ought to fix me," he wrote to Henslow in October 1836.[87] But soon (in November) he was telling Fox that ". . . all my affairs . . . are most prosperous; I find there are plenty who will undertake the description of whole tribes of animals, of which I know nothing."[88]

In addition to Owen in anatomy and Lyell and Lonsdale in geology, Darwin got the well-known English entomologists Frederick Hope and John Westwood to look at his insect specimens; Henslow and the prominent botanist Robert Brown he secured for botany; Dr. Grant for the corallines; Thomas Bell, Professor of Zoology at King's College, London, for reptiles; and the well-known conchologist William Broderip for the South American shells.

Also at this time, Darwin was attempting to put together, on a uniform plan or scheme, a volume dealing with the zoological results of the voyage. Darwin would be editor, with others (experts/ "lions") discussing the various categories of the specimens collected —birds, fish, and others. The idea was first suggested to him by several London zoologists. Darwin was intrigued. He thought it would be nice, he told Leonard Jenyns in almost a perfect description of his "collector" role on the *Beagle*, "to see the gleanings of my hands, after having passed through the brains of other naturalists, collected together in one work." He was also frank about the limited role he would play in the effort: "I myself should have little to do with it; excepting in some orders adding habits and ranges, & c., and geographical sketches, and perhaps afterwards some descriptions of invertebrate animals. . . ."[89]

In August 1837, through Henslow's help and Darwin's strenuous and time-consuming lobbying efforts, the project was approved and government money granted. In the autumn, Darwin spent several months working out the details of how the work would be organized.[90]

At the same time that he was arranging to have his specimens examined and trying to get the zoology of the *Beagle* work off the ground (which included recruiting experts for the various zoological categories), he was also attempting to prepare for publication his *Beagle* Journal and was hoping to begin work on what

would turn out to be several volumes on the geological results of
the voyage. The *Beagle* Journal and geology volumes were part of
an agreement Darwin had made with FitzRoy: the latter would
relate in two volumes the survey voyages of the *Beagle*—the latest one under FitzRoy, and the earlier one under Captain King—
and Darwin would prepare his journal as a third general volume,
which would be "a kind of a journal of a naturalist." Afterward,
Darwin would write an additional volume or volumes discussing
the geological results of the voyage in detail.[91]

But Darwin had problems. He could not finish the journal because there were small but important details that needed to be
filled in—details which Darwin could not supply because of his
lack of experience in several natural history areas. He sought help
from others.[92]

With regard to botany, for example, he wrote to Henslow in
May 1837: "There are about half a dozen plants of which if I do
not know the names of genus or something about them, I must
strike out long passages in my journal."[93] And again, in July, to
Henslow: "I am now hard at word, cramming up learning to orna-
ment my journal with; you may guess the object of this letter is
to beg a few hard names, respecting my plants . . . —Will you look
over the list of questions, & try to answer me some of them.—I
know if possible you will answer the questions."[94]

By August, Darwin was becoming anxious. Apparently Henslow, who had in the meantime agreed to read the proof sheets of
the Journal, had not yet replied to Darwin's list of questions. A
seemingly desperate Darwin supplicated his former Professor:
"Pray write *soon* & tell me whether you can answer me any of
the questions; so that I may know. . . . In less than a fortnight I
hope to send you my first proof sheet. . . . "[95]

Earlier (in May) Darwin had even thought of abandoning the
Journal for the time, feeling that he would be much better prepared from the standpoint of knowledge to write the Journal after
having completed the detailed work that would be involved in
putting together descriptions of the geology and zoology of the
voyage. As he wrote to Henslow: "I suspect I have begun at the
wrong end. I ought to have published detailed Geology & Zoology
first, & then all general views might have come out in as perfect
a form as the subject permitted.—Now the first book will consist
of mere series of imperfect sketches."[96] Darwin was able to com-

plete his journal by the end of 1837, though it was not published until 1839 because of delays by FitzRoy.

As if Darwin did not have enough to do, in the fall of 1837, he was asked to become one of the honorary secretaries of the Geological Society of London. In a letter to Henslow on October 14, 1837, he related the reasons why he was unwilling to accept the position. This letter is revealing with regard to the sort of work Darwin was then doing and the large amount of time it consumed, the work he felt he needed to do in the future, and the early stage at which he then perceived himself to be in his natural history career. At the outset, he lists his ignorance of English geology, his ignorance of all languages, and his inability "to pronounce a *single* word of French" as significant disqualifications for the job. In addition, he felt the loss of time the position might involve could be very serious: "Pray consider, that I should have to look after the artists, superintend and furnish materials for the government work [the zoology]. . . . All my geological notes are in a very rough state, none of my fossil shells worked up, and I have much to read. I have had high hopes by giving up society and not wasting an hour, that I should be able to finish my geology in a year & a half. . . . If this plan fails . . . the geology would necessarily be deferred till probably at least three years from this time . . . a great part of the utility of the little I have done would be lost, and all freshness and pleasure quite taken from me. . . . Moreover," he continued, "so EARLY in my scientific life, with so very much as I have to learn, the office, though no doubt a great honour, & c., could be the more burdensome."[97] Darwin eventually accepted the position and served in it three years, from February 16, 1838 to February 19, 1841. I have found no indication of how much time his duties actually consumed.

Amidst all his other activities during this crowded post-*Beagle* period, Darwin began collecting, as a side interest and hobby, facts bearing on the question of species. In July 1837, he opened his first notebook on species. Between that date and July 1839, Darwin filled four volumes of notebooks on this subject. Between July 15, 1838 and April 28, 1840, he filled an additional two notebooks with information on species and other topics—his so-called M and N notebooks. By the middle of September 1838, several weeks before reading Malthus, Darwin was writing to Lyell telling of his new, nongeological interests: "I have lately been sadly tempted to be idle—that is, as far as pure geology is concerned—

by the delightful number of new views which have been coming
in thickly and steadily,—on the classification and affinities and
instincts of animals—bearing on the question of species. Notebook after note-book has been filled with facts which begin to
group themselves *clearly* under "sub-laws."[98]

The notebooks provide yet another indication of the limited
state of Darwin's knowledge and experience at this time. Although they show very clearly the rapidity with which Darwin
was beginning to assimilate a vast storehouse of information, the
notebooks in both the pre- and post-Malthusian periods (July 1837
to April 1840) suggest the rudimentary state of Darwin's learning in these areas. The notebooks are replete with Darwin's personal instructions for further reading and research—the sort of
instructions one first entering an area of study would make: "Prove
animals like plants" (Transmutation Notebooks, I,73); "Consult
Dr. Smith History of S. African cattle" (I,167); "show independency of shells to external features of *land*" (II,99); "Old Buffon
should be read on mare" (III,112); "Buckland's Reliqu: Diluv. says
Africa only place where Elephant, Rhinoceros, Hippot., Hyaena
& are found together.—Read this Work.—" (IV,60)[99]

The notebooks seem as much a research agenda and bibliography as a focus for the discussion of species issues. And it is significant that they did not cease in October 1838 after Darwin had
read Malthus. They reveal Darwin in the process of learning—of
questioning, of collecting facts, of testing ideas. They reveal a mind
with limited knowledge seeking greater knowledge. In effect, they
stand as an important prelude to Darwin's research on the species question.

A Twenty-Year Process

What lay before Darwin, in terms of his species work, was the
enormous task of gathering information and learning the basic
precepts of a large number of natural history fields quite foreign
to him at the time, including classification, domestic breeding,
anatomy, physiology, hybridism, morphology, embryology, geographical distribution, instincts, heredity. He was beginning a
twenty-year process whereby he would attempt to match his creativity with knowledge in order to provide it with better direction
and support; and the process would be anything but simple, anything but easy.[100]

3

Lyell and the End of Geology

Working on Species

I have found no evidence to suggest that Darwin realized the state of his knowledge at the end of 1838 and sat down and devised a master plan to acquire the needed background in natural history. If Darwin did have a master plan, it was a poor one; his path to knowledge was, at best, meandering. Darwin sought knowledge when specific pragmatic needs required it. If he had a plan, it consisted of nothing more than a desire to read everything and collect all the facts he could about the question of species.

Between 1838 and 1854, Darwin's work on species was sporadic. There were long periods during this time—especially after 1844—when he did not seem to have worked directly on the problem at all. It was only after 1854 that his species work became a full-time occupation. Even before 1844, when he did considerable work, references to his species activities are sketchy. In Darwin's *Journal* or *Diary*, in which he kept a running account of his activities—researches, travels, illnesses, domestic affairs—no pattern of intense activity emerges, even between 1838 and 1844. In March 1839, Darwin mentions doing "a little work on Species"; in April, "some reading connected with Species"; in July and August, "read little for Species"; between December 24, 1839 and February 24, 1840, he "read a little for transmut. theory." In the summer of 1840, we see the first mention of active work on species: "when well enough did a good deal of Species work." In 1841, there is only one reference, in March, to species: ". . . sorted papers on Species theory." In May 1842, Darwin did some reading, and in June of the same year he "wrote pencil sketch of my Species theory." In 1843, there is again only one reference, in July: "some species work." In 1844, Darwin completed enlargement and improvement of the 1842 pencil sketch. In July 1844, he sent the enlarged sketch to be copied, and in September he made his final corrections on it. September 1844 is the last reference in the *Diary* to species work prior to 1854.[1]

It would be dangerous, however, to equate Darwin's intermittent work on his species theory with any lack of interest in it. We get an indication of the importance he attached to the subject in a letter he wrote to his wife, Emma, soon after completing his enlarged species sketch in 1844. In this letter he provided instructions for the disposition of his theory in the case of his untimely death:

I have just finished my sketch of my species theory [Darwin wrote]. If, as I believe, my theory in time be accepted even by one competent judge, it will be a considerable step in science. I therefore write this in case of my sudden death, as my most solemn and last request, which I am sure you will consider the same as if legally entered in my will, that you will devote £400 to its publication, and further, will yourself, or through Hensleigh [Emma Darwin's brother, Hensleigh Wedgwood], take trouble in promoting it. I wish that my sketch be given to some competent person, with this sum to induce him to take trouble in its improvement and enlargement.

Darwin then provided some instructions as to how the editor might be assisted, what materials might be available to him, and who might prove a suitable editor.

With respect to editors, Mr. Lyell would be the best if he would undertake it; I believe he would find the work pleasant, and he would learn some facts new to him. As the editor must be a geologist as well as a naturalist, the next best editor would be Professor Forbes of London. The next best (and quite best in many respects) would be Professor Henslow. Dr. Hooker would be *very* good [emphasis Darwin's]. . . . If none of these would undertake it, I would request you to consult with Mr. Lyell, or some other capable man for some editor, a geologist and naturalist.

Ten years later, Darwin added to the back of this letter: "Hooker by far best man to edit my species volume. August 1854."[2]

Darwin clearly thought his theory important enough to safeguard for posterity—that his theory be preserved was his "most solemn and last request" to be considered a legal part of his will. He was at this time (1844) often ill, and his fear of possible sudden death was not unreasonable. But why, if the theory was so important to him, did he not turn full attention to it in 1844 or even earlier? Why did he wait until 1854 to begin a full exposition of it?

I do not think that Darwin ever completely put his species work aside. By the end of 1838, Darwin had a tentative species theory worked out. Darwin's problem at that time, as I have stressed, was that his theory had, in effect, preceded his knowledge, and he needed to get caught up. He spent the years 1838–54 doing just that—not always by direct work on species, but often by work in closely associated fields: geology, zoology, paleontology.

Between 1846 and 1854, for example, Darwin was occupied full-time with a study of the Cirripedes. One reason for this study was Darwin's desire to gain experience in biology—experience he felt he needed for his species work. As he wrote to Hooker in September 1845: "All of what you kindly say about my species work does not alter one iota my long self-acknowledged presumption in accumulating facts and speculating on the subject of variation, without having worked out my due share of species."[3] Thomas Henry Huxley later applauded Darwin's work on Cirripedes and the foresight it showed regarding Darwin's need for experience in biology. As he wrote to Darwin's son, Francis: "In my opinion your sagacious father never did a wiser thing than when he devoted himself to the years of patient toil which the Cirripede-book cost him. Like the rest of us, he had no proper training in biological science, and it has always struck me as a remarkable instance of scientific insight, that he saw the necessity of giving himself such training, and of his courage, that he did not shirk the labour of obtaining it."[4] Thus, Darwin's Cirripedes work, to which he devoted eight years, did relate indirectly to his study of species.

Also, Darwin had intended to work on Cirripedes for only a short period before beginning work on species in earnest. "I hope this next summer to finish my South American Geology," he wrote to Hooker in 1845, "then to get out a little Zoology [i.e., Cirripedes], and hurrah for my species work. . . . "[5] Darwin spent the next eight years "getting out a little zoology." Two of these years, according to Darwin, were lost to illness.[6] That leaves six years to be accounted for. Obviously, the "little zoology" expanded its scope. Part of the problem was Darwin's approach to the subject. He found it impossible to treat any subject, however circumscribed, in a superficial manner. Darwin was a slow, methodical, painstakingly thorough worker. Many of his works, as a result, took a great deal of time to complete. His work on Cirripedes was no exception.[7]

Before 1846, the intermittent nature of Darwin's work on his species theory was really a question of other priorities. Following his return from the *Beagle* voyage, Darwin had, as we have seen, an enormous amount of work to do: disposal of his specimens, preparation of the zoological and geological volumes of the *Beagle* voyage, the writing of his *Journal of Researches*, performing duties as secretary of the Geological Society. Prior to 1846, Darwin could not work full-time on his species theory because he simply had too many other things to do.

In addition, during this same period Darwin had embarked on a successful career in geology. In this field he was fast becoming a "lion" in his own right. There was no need between 1836 and 1846 for Darwin to turn to other areas for recognition of his talents. His species theory during the same period, though undoubtedly important to him, was, as we have seen, a far-off thing. Darwin chose a burgeoning career over an uncertain theory. To abandon all else for species at the time would have been foolish.

Let us look in more detail at Darwin's career in geology, which dominated the ten years following his return from the *Beagle* voyage, and the reasons for its initial success, and eventual demise.

Charles Lyell and Uniformitarianism

Charles Lyell is the key person in this period of Darwin's career.[8] When Lyell published the first volume of his *Principles of Geology* in 1830,[9] he was attempting to provide for geology a comprehensive theory to account for all possible past and present geological changes. He wanted to establish that there was an "undeviating conformity of secondary causes" guiding geological change—that the forces working to transform the surface of the earth in the past were the same as those that could be seen in operation at present, and that these forces were ordinary, regular, orderly, and lawlike. Lyell eschewed the supernatural or spiritual origin of geological processes.

The net effect of Lyell's uniform, lawlike forces was slow, non-directional, cyclical change occurring over vast periods of time. Elevation was balanced by subsidence, deposition by erosion. And although the earth was in a state of constant flux, it was not moving in any particular direction or towards any particular goal. Mountains might decay and new mountains be thrust up again, but these new mountains were not considered to be more complex or very different from previous ones. For Lyell, there were no basic changes in the forms of natural phenomena.

To emphasize his point about nondirectional, cyclical change in the inorganic world, Lyell sought confirmation in the organic. He had another motive. In the organic world some had claimed that a progression of forms—a movement from the very simple to the very complex—could be seen. Lyell feared that an admission of progress in the organic world would imply similar progress in the inorganic. Because of changing geological and climatic environments, and because of the evidence of fossil remains, Lyell was willing to admit that certain species had become extinct and that new species, by whatever means, were being created, but he felt that new species were not any better or organizationally higher than previous ones, and, thus, that progress in the organic world did not exist. In an exposition and critical analysis of the French naturalist J. B. Lamarck's evolutionary views in Volume II of the *Principles*,[10] Lyell concluded that evolution had not occurred, and that it could not. Because of the struggle for existence among organisms—a struggle which seemed to be present everywhere in nature—organisms did not have time to develop new organs in response to changes in the environment, as envisioned by Lamarck. Evolution could not occur because the struggle for existence was both too harsh and too immediate.

Thus, in the organic world, as in the inorganic, there was change, there was movement, there was destruction and birth, but there was no real progress toward any particular goal.[11]

One of the most popular contemporary views of geologic change which Lyell was seeking to combat was that which saw the earth, through a succession of violent and cataclysmic events, moving in interrupted though progressive steps towards its present state. Proponents of this view felt that geological forces in the past were distinct in kind and in degree from forces in operation at present, and that these forces were often of an extraordinary, irregular, and chaotic nature. They envisioned directional, progressive change over relatively short periods of time, and attributed much of this change to the force of spiritual powers, such as those causing the Biblical Flood. Unlike Lyell, they saw progress in the organic world, where fossil remains seemed to extend in a never-ending progression from the oldest to the most recent geological strata.

A lively debate between advocates of both views of geological development directly followed the publication of Lyell's work. The Uniformitarian-Catastrophist debate, as the controversy between these groups has been called, was not so much a debate between progressive science and religious orthodoxy as it was a debate within science itself (or, more accurately, within the Geological Society of London). On some of the most important issues Uniformitarians and Catastrophists were often not far apart.[12] Lyell, for example, did not deny the Flood, he merely put it in the context of other natural processes occurring throughout earth history. He stressed the catastrophic nature of certain geological processes. Also, Lyell did not openly confront religious authority. On the contrary, he was cautious in his *Principles* not to offend religious sensibilities, and he lobbied hard before the publication of his work for support from potential religious adversaries. He himself excluded man from the operation of secondary causes, attributing man's creation to divine intervention; he felt that to argue the Uniformitarian cause was to affirm rather than to deny the importance of the spiritual in geological development.[13] Lyell's only concern was that such a question was not one within the proper scope of science.

Most Catastrophists, for their part, were not fire-eating religious fanatics, at least when it came to geological matters. For a number of years they had liberalized their views of the Biblical Flood, bringing its effects within what seemed to be geologic reason. Adam Sedgwick, one of the leading Catastrophists, had repudiated the view of the Flood as a primary and universal geological event even before Lyell's work had appeared. In Sedgwick's presidential address to the London Geological Society in 1830, he attacked "the unnatural union of things so utterly incongruous" as science and revelation. He felt that science had to pursue its goals independent of the Bible.[14] And William Whewell, another leading Catastrophist and Professor of mineralogy at Cambridge, found Lyell's theories unsupported by known geological evidence, yet felt that the "hammer-bearing" Lyell had produced a work of "great ingenuity" and "great boldness," which was "a skillful and masterly attempt to combine into a consistent view a large mass of singularly curious observations and details, which no one but an accomplished geologist could have brought together." Whewell felt that the time had come when intelligent people would look without suspicion at the sort of theories Lyell had offered and that "the condition and history of the earth, so far as they are independent of the history and condition of man, are left where they ought to be, in the hands of the natural philosopher. . . ."[15]

The real enemies of both Uniformitarians and Catastrophists were the religious and pseudoreligious zealots who flailed against both groups, though it would be a distortion to imply there were not important differences between Uniformitarian and Catastrophist positions.

Meeting Lyell

Lyell's *Principles of Geology* had a large immediate sale; new editions of the first and second volumes were needed before the third was even published. Before Darwin departed on the *Beagle* voyage, Henslow provided him with a copy of Lyell's first volume and advised him to read it but by no means to believe it. Henslow, like Adam Sedgwick, was a Catastrophist. Darwin received the second and third volumes of the *Principles* later in the voyage. He probably read the first volume between the time the *Beagle* departed from Portsmouth and the time it arrived in St. Jago.[16]

Geology, as we saw in the last chapter, became Darwin's abiding interest during the *Beagle* voyage, and he had found in Lyell's *Principles* an indispensable guide in interpreting the geology of the various places visited during the voyage. Upon Darwin's return to England in 1836, Lyell, aware of Darwin's work abroad, sought him out, and a strong friendship—professional and social—soon developed. "Mr. Lyell has entered in the *most* goodnatured manner, & almost without being asked, into all my plans . . .," Darwin wrote to Henslow on October 30, 1836; "you cannot conceive anything more thoroughly goodnatured, than the heart & soul manner, in which he puts himself in my place & thought what would be best to do. . . ."[17] Some six months later, on April 10, 1837, Darwin revealed to Leonard Jenyns that he had "a capital friend in Lyell, and see a great deal of him . . . ";[18] and to Fox, in July of 1837, he confessed: "I never expected that my Geology would ever have been worth the consideration of such men as Lyell, who has been to me, since my return, a most active friend."[19]

Lyell seemed equally happy with the young Darwin. "It is rare," Lyell wrote to Adam Sedgwick on April 25, 1837, "even in one's own pursuits to meet with congenial souls & Darwin is a glorious addition to my society of geologists & is working hard & making way both in his book and in our discussions." Lyell especially liked Darwin's spunk—the way, he told Sedgwick, Darwin deftly handled a pesky questioner who had asked "some impertinent & irrelevent questions about [the geology of] S. America" at a Geological Society meeting. "I never really saw that bore Dr. Mitchell [the questioner] so successfully silenced or such a bucket of cold water so dexterously poured down his back," Lyell gleefully reported.[20]

With time, Darwin and Lyell grew very close. They frequently dined together, read and criticized each other's works (often in draft or proof form), and were frequently guests at each other's homes. Perhaps the closeness of their friendship became most visible when the prospect arose that the two men might be separated for some time. In 1841, Lyell was about to leave for a year's tour of North America and Darwin was preparing to move away from London to the country, to Down House in Kent. In July 1841, Darwin had gone to Shrewsbury to visit his father and was away when Lyell was readying to depart on his voyage. Lyell wrote to him at the time: "We were much disappointed in not seeing you before your start for a year's absence. I cannot tell you how often since your long illness I have missed the friendly intercourse which we had so frequently before, and on which I built more than ever after your marriage. . . . The prospect of your residence somewhat far off, is a privation—I feel as a very great one—I hope you will not, like Herschel, get far off from a railway."[21]

In 1845, Lyell was again preparing to leave for abroad. Darwin wrote to him in July of that year: "How sorry I am to think that we shall not see you here again for so long; I wish you may knock yourself a little bit up before you start and require a day's fresh air [at Down], before the ocean breezes blow on you . . . ";[22] and again in October: "We shall miss . . . your visits to Down, and I shall feel a lost man in London without my morning 'house of call' at Hart Street [Lyell's London residence]. . . . "[23]

Sources of the Friendship

What were the sources of this close friendship? I see several factors drawing the two together. From Darwin's point of view, Lyell had provided him, soon after his return from the *Beagle* voyage, with an introduction to the London scientific community. He had welcomed Darwin enthusiastically to the Geological Society, of which Lyell was president in 1836. One of his last acts as president in that year was to nominate young Darwin to the presiding council of the Society. I suspect that through dinners and teas at the Lyells' Darwin got to meet many of the most prominent and aspiring scientific men of the day. The Lyells were socially gregarious. At one such tea party in Lyell's home in October 1836, Darwin met young Richard Owen, whom Lyell had invited to the party with the added inducement that "you will meet Mr. Charles Darwin."[24]

Also, Lyell showed great interest in, and support for, Darwin's geological discoveries, offering Darwin important private encouragement and public recognition of his work. In 1835, Adam Sedgwick had presented to a meeting of the Geological Society an acount of Darwin's geological explorations in South America and coastal West Africa. Sedgwick extracted his account from letters Darwin was sending back to Henslow during the *Beagle* voyage. In his presentation, Sedgwick described the substance of Darwin's geological work: the accumulation of evidence for the recent elevation of the Island of St. Jago, in the Cape Verde Islands chain off the coast of West Africa; evidence for the elevation of the Southern Andes before the Tertiary period; descriptions of the Tertiary groups on both sides of the chain of the Andes; descriptions of a transverse section of the Cordillera of the Andes; evidence for the gradual elevation of the Andes owing to a succession of small elevations, like those experienced during that time in Chile; separation of the Tertiary deposits of Patagonia into distinct periods "somewhat similar to those derived by Mr. Lyell from a comparision of the newer deposits of Europe." In addition to "these very remarkable notices," Sedgwick mentioned the discovery of many fossil bones of large, extinct quadrupeds found near the banks of the Rio Plata, in the Pampas of Buenos Aires, and in the gravel of Patagonia.[25]

Lyell, in his February 19, 1836 presidential address before the Society, noted that "few communications have excited more interest in the Society than the letters on South America addressed by Mr. Charles Darwin to Professor Henslow." Lyell then described Darwin's discoveries.[26] This was eight months before the *Beagle* brought Darwin back to England.

On October 2, 1836, the *Beagle* dropped anchor in Falmouth harbor. On November 30, 1836, Darwin was elected a fellow of the Geological Society. On January 4, 1837, Darwin presented before the Society a paper entitled "Observations of proofs of recent elevation on the coast of Chile."[27] Darwin had sent a draft of this paper to Lyell in late December 1836. They evidently discussed the paper on January 2, 1837, when Darwin was at the Lyells' for dinner.[28] In this paper Darwin sought to provide proof of the recent elevation of the coastline of Chile owing to a major earthquake of 1822. He stressed that such sudden elevations were independent of larger, slower-working geological forces that were insensibly and gradually raising the coast of Chile over geologic ages. This observation was in direct support of Lyell's Uniformitarian theories, with their emphasis on slow, gradual, noncatastrophic change.

On February 17, 1837, Lyell delivered his second presidential address to the Society. As in the 1836 address, Darwin was prominently mentioned. "While a variety of geological monuments are annually discovered which attest modern alterations in the level of the land," Lyell pointed out, "it is important to remark that new testimony is also daily obtained of the rising and sinking of land in our own times. . . . Mr. Darwin, whose opportunities of investigation both in Chile and other parts of South America have been so extensive, thinks it quite certain that the land was upheaved two or three feet during the earthquake of 1822, and he met with none of the inhabitants who doubted the change of level."

"In Mr. Darwin's paper," Lyell added, "you will find other facts elucidating the rise of land . . . and he has treated of the general question of the elevation of the whole coast of the Pacific. . . . "[29]

Later in the address, Lyell mentioned Darwin's fossil discoveries. At the time, Lyell was communicating with Richard Owen, who was examining Darwin's fossils for the College of Surgeons, where Darwin had deposited them. Lyell seemed quite interested in Owen's research. In April 1837, Owen presented to the Society a description of the bones. Owen's presentation was followed on May 3, 1837, by a paper of Darwin's entitled "A Sketch of the Deposits containing extinct Mammalia in the neighbourhood of the Plata."[30] Lyell saw the significance of the fossil remains. The fossil discoveries meant, he told the Society, that "the peculiar type of organization which is now characteristic of the South American mammalia has been developed on that continent for a long period, sufficient at least to allow of the extinction of many large species of quadrupeds."[31] This was a theme that Darwin expanded upon in his May 3 paper.

At the same time Lyell was supporting Darwin's work on the elevation of South America and becoming interested in Darwin's South American fossil discoveries, he was fascinated by a new theory of Darwin's regarding the origin and development of coral reefs. Lyell's support for Darwin's work in this area is impressive, because Darwin's theory of coral reefs forced Lyell to abandon his own cherished view of their development.

In the *Principles of Geology*, Lyell had suggested that coral atolls represented reefs built on the rims of submerged volcanic craters.[32] By 1836, Lyell's theory had become generally accepted. Darwin, however, maintained that although Lyell's theory made some sense in terms of one sort of coral formation—namely, the lagoon coral reef—it did not help explain the origin of several other coral formations. For example, encircling reefs, which form a ring around mountainous islands, are at a distance of two to three miles from the shore, rise from a profoundly deep sea, and are separated from the land by a deep channel. They could not be accounted for by assuming they were developed on external craters. Nor could Lyell's theory explain the existence of the great Australian barrier reef, which runs for nearly a thousand miles parallel to the northeast coast of that continent and encompasses a wide and deep area of the sea.

Darwin's theory was based, in good Uniformitarian fashion, on a supposition of massive but slow subterranean subsidence. Darwin suggested that inasmuch as coral cannot grow in very deep water, it would tend to build near the shores of islands. As these islands, together with their closely connected reefs, subsided gradually, the coral building polypi would gradually raise the level of the coral to sea level. But the land would continue to sink. Eventually, the land would disappear altogether, and only the coral would remain as atolls of one form or another, depending on the configuration of the subsided land.[33]

After hearing Darwin's theory, Lyell "was so overcome with delight that he danced about and threw himself into the wildest contortions, as was his manner when excessively pleased."[34] On May 24, 1837, he wrote to his friend, the astronomer John Herschel: "I am very full of Darwin's new theory of Coral Islands, and have urged Whewell [who had been elected the new President of the Geological Society] to make him read it at our next meeting. I must give up my volcanic crater theory for ever, though it costs me a pang at first, for it accounted for so much. . . . "[35] And to Darwin, who did not realize the significance of his theory until Lyell pointed it out to him, he wrote: "I could think of nothing for days after your lesson on coral reefs, but of the tops of submerged continents. It is all true, but do not flatter yourself that you will be believed till you are growing bald like me, with hard work and vexation at the incredulity of the world."[36] On May 31, 1837, Darwin presented his coral reefs theory to the Society.[37]

Another factor drawing Darwin toward Lyell seems to have been the pure intellectual attraction of Lyell's theories. To Darwin, Lyell's geology was by far the best available method for interpreting geological phenomena. "After having just come back from Glen Roy, [I have] . . . found how difficulties smooth away under your principles," he wrote to Lyell on September 13, 1838.[38] And to his fellow geologist, Leonard Horner, on August 29, 1844, Darwin admitted: "I cannot say how forcibly impressed I am with the infinite superiority of the Lyellian school of Geology over the continental."[39] Darwin was pleased to hear the call in 1845 for yet another edition of the *Principles:* "What glorious good that work has done," he wrote to Lyell.[40]

Darwin was never completely satisfied that he had adequately paid his geological debt to Lyell. In his letter to Horner, in August

1844, cited above, Darwin explained that he always felt that his books "came half out of Lyell's brain," and that he had never acknowledged this sufficiently. "Nor do I know how I can," he added, "without saying so in so many words—for I have always thought that the great merit of the *Principles* was that it altered the whole tone of one's mind, and therefore that, when seeing a thing never seen by Lyell, one yet saw it partially through his eyes—it would have been in some respects better if I had done this less."

In 1845, Darwin dedicated the second edition of his *Journal of Researches* to Lyell. The inscription read: "To Charles Lyell, Esq., F.R.S., this second edition is dedicated with grateful pleasure—as an acknowledgement that the chief part of whatever scientific merit this Journal and the other works of the Author may possess, has been derived from studying the well-known and admirable 'Principles of Geology.' "[41] In July of that year, Darwin wrote to Lyell: "I have long wished, not so much for your sake, as for my own feelings of honesty, to acknowledge more plainly than by mere reference, how much I geolgically owe you. Those authors . . . who like you, educate people's minds as well as teach them special facts, can never, I should think, have full justice done them except by posterity, for the mind thus insensibly improved can hardly perceive its own upward ascent."[42]

Finally, Darwin had found in Lyell a compassionate and sympathetic advisor and hero figure, someone he could feel comfortable with, yet at the same time admire. Darwin was especially impressed with Lyell's broadmindedness and sympathy to new ideas. "I mean to have a good hour's enjoyment and scribble away to you, who have so much geological sympathy that I do not care how egotistically I write," Darwin wrote half-jokingly to Lyell in 1838.[43] And on March 9, 1841, Darwin confessed to his friend: "It is the greatest pleasure to me to write or talk Geology with you."[44]

Like Henslow before him, Lyell assumed the role of counselor and advisor to his young protégé, and Darwin seemed to delight in Lyell's advice and direction, which he found to be unerring. We have already seen Darwin indicate to Henslow that Lyell had entered into, soon after Darwin's return from the *Beagle* voyage, Darwin's plans for his future. Lyell offered advice to Darwin on numerous other subjects as well: on accepting official positions— "Don't accept any official scientific place, if you can avoid it, and

tell no one that I gave you this advice";[45] on how to furnish a
home—slowly, because it would be more economical that way;[46]
on working hard, but not too hard—"I have every motive to work
hard," Darwin wrote Lyell, "and will, following your steps, work
just that degree of hardness to keep well";[47] on the ability to get
away from thoughts of one's work;[48] on cutting one's working day
into halves, thus breaking the monotony of long hours of study
and research;[49] on the advisability of attending British Association for the Advancement of Science meetings, although Lyell
himself ("your advisor" as he described himself to Darwin) had
to admit that he had only attended two of the previous eight held.[50]

On August 9, 1838, Darwin wrote to his new-found guardian:
"I have come to one conclusion, which you will think proves me
to be a very sensible man, namely, that whatever you say proves
right."[51]

Lyell had found in Darwin some one personally agreeable, supportive, intelligent, and intellectually compatible. An added attraction from Lyell's perspective was the leisure that Darwin's
financial independence allowed him. As Lyell wrote to Darwin:
"I heartily long for some one here with a collection of shells, and
leisure to talk on these matters with. Lonsdale is overpowered
with work."[52] Lyell also liked Darwin's unstinting honesty and
self-effacement. Darwin, Lyell wrote to his sister-in-law, Susan
Horner, was "the most candid of men & if anything is new, says
at once he never had thought of it."[53] But most important to Lyell
was the support Darwin was giving his theories—support based
on new and impressive geolgical evidence brought back from half-
way round the world. Lyell needed such support at the time; he
was well aware that many of his contemporaries still viewed Uniformitarian theories with suspicion.

Lyell had agreed to review Darwin's paper on the gradual elevation of the Andes toward the end of 1836 before Darwin's presentation of it to the Geological Society in January 1837. Lyell
wrote to his young friend on December 26, 1836: "I have read
your paper with the greatest pleasure, and should like to point
out several passages which require explanation . . . the idea of the
Pampas going up, at the rate of an inch in a century, while the
Western Coast and Andes rise many feet and unequally, has long
been a dream of mine. What a splendid field you have to write
upon!"[54]

At a meeting of the Geological Society on March 7, 1838, Darwin read a paper in which he argued that earthquakes and volcanic action were both manifestations of the internal forces that were elevating the continent of South America. Darwin was able to show that larger mountain chains, such as the Andes, could be elevated gradually. As he stated in Lyellian terms: "The important fact which appears to me proved, is that there is a power now in action, and which has been in action with the same average intensity . . . since the remotest periods, not only sufficient to produce, but which almost inevitably must have produced, unequal elevation on the lines of fracture."[55] This was, according to Leonard Wilson, the most extensive and strongly supported argument to be launched against the French geologist Elie de Beaumont's anti-Uniformitarian theory of the paroxysmal elevation of mountain chains, and it was based on an examination of a major mountain chain, the Andes.[56]

Lyell was impressed. He saw in the discussion that followed the presentation of Darwin's paper an important favorable shift in opinion regarding his theories. As he wrote to Leonard Horner on March 12, 1838: "About the last meeting of the G.S. where Darwin read a paper on the connexion of volcanic phenomena & elevation of mountain chains in support of my heretical doctrines . . . I was much struck [in the discussion which followed Darwin's presentation] with the different tone in which my gradual causes were treated by all, even including de la Bêche from that which they experienced in the same room 4 years ago when Buckland, de la Bêche, Sedgwick, Whewell and some others [all Catastrophists] treated them with as much ridicule as was consistent with politeness in my presence." In a postscript added a day or two later, Lyell revealed that Darwin had felt very differently, more negatively, about the discussion because, according to Lyell, Darwin had not been able to measure the change in tone over the past few years. Lyell, however, was confident that he had restored Darwin "to an opinion of the growing progress of the true cause."[57]

Lyell was certainly grateful for Darwin's support on this issue. In a letter to Darwin, written on September 6, 1838, Lyell suggested that Darwin's "grand discovery" of the elevation of the South American continent had proved "in the most striking manner, the weight of my principal objection to the argument of De Beaumont." And in a postscript two days later, Lyell added a hopeful note: "I really find, when bringing up my Preliminary Essays in 'Principles' to the science of the present day . . . that the great outline, and even most of the details, stand so uninjured, and in many cases they are so much strengthened by new discoveries, especially by yours, that we may begin to hope that the great principles there insisted on will stand the test of new discoveries."[58]

Darwin, sensing Lyell's new enthusiasm, added his own optimistic appraisal. As he wrote to Lyell on Spetember 13, 1838: "You say you 'begin to hope that the great principles there insisted on will stand the test of time.' *Begin to hope:* Why, the *possibility* of a doubt has never crossed my mind for many a day. This may be very unphilosophical, but my geological salvation is staked on it . . . it makes me quite indignant that you should talk of *hoping* . . . [that your principles will be eventually accepted]."[59]

Abandoning Geology

By 1841 or 1842, armed with new and exciting geological evidence obtained in South America and bolstered by the full support of one of the most prominent geologists of the day, Darwin was beginning to forge a promising career in geology. He had presented a number of papers before the Geological Society of London, several of which had excited considerable interest. His paper, "Observations on the Parallel Roads of Glen Roy," had appeared in the prestigious *Transactions* of the Royal Society. He had published his *Journal of Researches* and three substantial volumes on the geological results of the *Beagle* voyage. And he had gained prominent attention as a member of the Council of the Geological Society and as one of its honorary secretaries.

Yet, by 1846, Darwin had, for all intents and purposes, turned from geology and plunged into a tedious zoological study which would occupy him for the next eight years, with only brief diversions into geology and other fields.

What had happened?

One thing that happened was that after 1842 Darwin was no longer living in London. His visits to Lyell in Hart Street for geological discussions did not occur as frequently as before, nor could Darwin attend Geological Society meetings as often as before. Did distance breed indifference? It may have contributed to it.

Also, Emma Wedgwood, whom Darwin married in 1839, did not seem, at least at one point, particularly fond of Lyell. "Mr. Lyell is enough to flatten any party," Emma wrote to her sister soon after the newly married couple had given their first London dinner party on April 1, 1839; "he never speaks above his breath, so that everybody keeps lowering their tone to his."[60] Whether Emma's initial irritation with Lyell continued, I cannot tell. Lyell's July 1841 letter to Darwin, cited above, does suggest that he had high hopes of seeing Darwin even more frequently after his marriage to Emma, but that his hopes had not been realized.

Another problem arose because of Darwin's increasingly poor health. Darwin showed signs of illness before the *Beagle* voyage when he was waiting at Portsmouth harbor for the *Beagle* to depart. During the voyage itself, Darwin experienced several long illnesses and was constantly suffering from seasickness, though apparently his energy was not adversely affected. After his return to England in 1836, Darwin became increasingly ill for longer periods of time. In June 1838, Darwin was preparing to depart on his geological excursion to Glen Roy, Scotland. He wrote gloomily to Fox: "I have not been very well of late, which has suddenly determined me to leave London earlier than I anticipated . . . I intend stopping a week to geologise the parallel roads of Glen Roy, thence to Shrewsbury, Maer for one day, and London for smoke, ill health and hard work."[61] Toward the end of 1839, Darwin felt compelled to give up London social events because of the strain they produced on his health. In February 1840, a forlorn Darwin wrote to Lyell: "Is it not mortifying, it is now nine weeks since I have done a whole day's work, and not more than four half days."[62] And to Fox (September 1841), Darwin lamented: "I grow very tired in the evenings, and am not able to go out at that time, or hardly to receive my nearest relations. . . ."[63]

Darwin was fearful that his poor health would severely limit the contribution he could make to natural history. "It has been a bitter mortification for me to digest the conclusion that the 'race is for the strong,' " he wrote pessimistically to Lyell in June 1841, "and that I shall probably do little more but be content to admire the strides others make in science."[64]

By 1842, Darwin had suffered over a dozen serious illnesses: lassitude, gastrointestinal discomfort, nausea, prolonged periods of vomiting, and sleeplessness were among the more pronounced symptoms. In September 1842, his health deteriorating, Darwin and his wife moved to semiseclusion at Down. For the next forty years, Darwin led a semiinvalid existence.

A glance at the undeviating daily routine that gradually evolved at Down should give some indication of how severely debilitating Darwin's ill health was. His son, Francis, has provided us a description of this routine. Darwin rose early, "chiefly because he could not lie in bed," and took a short walk before breakfasting alone at about seven forty-five. He worked from eight until nine-thirty, when he entered the drawing room for the daily mail. "He would then hear any family letters read aloud as he lay on the sofa." The "reading aloud" was Emma's task, and "lasted till about half-past ten, when he went back to work till twelve or a quarter past." "By this time he considered his day's work over, and would say, in a satisfied voice, '*I've* done a good day's work.' " He then went out for his midday walk. Lunch followed and then, "lying on the sofa in the drawing-room," he read the daily newspaper. After the newspaper, there was time set aside for letter writing. This lasted until three, at which time he mounted the stairs to his bedroom, where he would lie on the sofa, smoke a cigarette, and listen to a novel "or other book not scientific." At four he came down to dress for his afternoon walk. "From about half-past four to half-past five he worked; then he came to the drawing room, and was idle till it was time (about six) to go up for another rest with novel-reading and a cigarette." Dinner was at half-past seven. He dreaded social intercourse after dinner, not because he did not enjoy conversation, but because of the ill effects —"a sleepless night" or "the loss perhaps of half of the next day's work"—such activity usually produced. The evening consisted of two regular games of backgammon with Mrs. Darwin, the enjoyment of a scientific book, often the appreciation of Emma at the piano. At about half-past ten he went off to bed, but often not to sleep. "His nights were generally bad and he often lay awake or sat up in bed for hours, suffering much discomfort. He was troubled at night by the activity of his thoughts, and would become exhausted by his mind working at some problem which he would willingly have dismissed. At night, too, anything which had vexed or troubled him in the day would haunt him. . . . "[65]

Any activity outside this daily routine was attempted with only great pain and difficulty. Public appearances Darwin found ex-

hausting; social visits, wearing; even holidays were agreed to reluctantly. Francis recalled: "[Father] . . . was generally persuaded
by my mother to take . . . short holidays, when it became clear
from the frequency of 'bad days,' or from the swimming of the
head, that he was being overworked. He went unwillingly, and
tried to drive hard bargains, stipulating, for instance, that he should
come home in five days instead of six. . . . The discomfort of a
journey to him was . . . chiefly in the anticipation, and in the miserable sinking feeling from which he suffered immediately before
the start. . . . "[66]

To be ill and attempt to continue any career would be difficult;
but to be ill, as Darwin was, and attempt to continue a career in
geology was almost impossible. Geology was not a sedentary profession. Darwin knew that in order to pursue an active geological career with the hope of original contribution, he would need
to travel extensively and live, at times, under harsh conditions,
as he had during his geological excursions in South America. Lyell,
Darwin could see, was always traveling on a geological expedition to some place near or far. "It makes me groan to think that
probably I shall never again have the exquisite pleasure of making out some new district, of evolving geological light out of some
troubled dark region," a distraught Darwin wrote to Lyell on September 14, 1849. "So I must make the best of my Cirripedia . . . ,"
he sadly concluded.[67]

Still another problem for Darwin was the lack of attention paid
to his geological works by the mid-1840s. This fact seemed particularly to irritate him. "I have long discovered that geologists
never read each other's works," Darwin wrote to his old college
friend, J. M. Herbert, in 1844 or 1845, "and that the only object
in writing a book is a proof of earnestness, and that you do not
form your opinions without undergoing labour of some kind.
Geology is at present very oral. . . . "[68]

To Leonard Horner he expressed similar disappointment. He
wrote on August 27, 1844: "But excuse this lengthy note and once
more let me thank you for the pleasure and encouragement you
have given me—which, together with Lyell's never-failing kindness, will help me on with [my] South America [geology], and, as
my books will not sell, I sometimes want such aid."[69]

Writing to Lyell in July 1845 concerning the dedication of the
second edition of Darwin's *Journal of Researches*, Darwin ex-

plained that he had intended to put the present acknowledgment in the third volume of his geology of the *Beagle* voyage, "but its sale is so exceedingly small that I should not have had the satisfaction of thinking that as far as lay in my power I had owned . . . my debt."[70]

One final example: in the spring of 1849, Hooker had written to Darwin requesting advice concerning the possible publication of some of Hooker's geological letters in the prominent British journal *Athenaeum*. Darwin replied that he did not think Hooker's letters were appropriate for that journal and added bitterly: "I have no interest; the beasts [the *Athenaeum*] not having even *noticed* my three geological volumes which I had sent to them. . . ."[71]

Yet another major problem arose from the sort of contribution to natural history Darwin had envisioned making. Darwin was, throughout his life, concerned with recognition and with achievement, even though at times, perhaps out of shame or guilt, he would disclaim any such interests. Even as early as 1818 at Dr. Butler's, all Darwin cared about was a "new *named* mineral."[72] At Edinburgh he regretted that papers submitted to the Plinian Society were not published and "I had not the satisfaction of seeing my paper in print. . . ."[73] At Cambridge his entomological interests gave him great pleasure when he saw recognition of his efforts in a leading British entomological journal. "No poet ever felt more delight at seeing his first poem published," he later remembered, "than I did at seeing in Stephan's *Illustrations of British Insects* the magic words, 'captured by C. Darwin, Esq.'"[74]

During his stay at Cambridge, as we have seen, Humboldt's and Herschel's writings stirred in him "a burning zeal to add even the most humble contribution to the noble structure of Natural Science." During the voyage of the *Beagle*, Darwin worked to "the utmost" from the mere pleasure of investigation and from his desire to add, as he later remembered, "a few facts to the great mass of facts in natural science. But I was also ambitious to take a fair place among scientific men," Darwin continued, "whether more ambitious or less so than most of my fellow-workers I can form no opinion."[75] At St. Jago, we will remember, Darwin first conceived the idea of writing a book on the geology of the various places visited during the voyage and this made him "thrill with delight." Later, when FitzRoy suggested that Darwin's journal would be worth publishing, Darwin was enthralled again with the prospect of authorship.

During the voyage, Darwin sensed the great potential that geology held as a field of study. "I find in Geology a never failing interest," we recall he wrote to his friend Whitley in 1834; "as it has been remarked, it creates the same grand ideas respecting this world which Astronomy does for the universe."[76] Toward the end of the voyage, Darwin received a letter at Ascension, in which his sister told him of a visit Sedgwick had made to Darwin's father. Sedgwick apparently predicted that Charles would someday take a place among the leading scientific men. "After reading this letter," Darwin later recalled, "I clambered over the mountains of Ascension with a bounding step and made the volcanic rocks resound under my geological hammer! All this shows how ambitious I was." But in later years, Darwin felt that he cared only for "the approbation of such men as Lyell and Hooker, who were my friends . . ., [and] did not care much about the general public."[77]

After Darwin's return to England, he was delighted with the attention he was getting. To Fox, on November 6, 1836, Darwin gleefully revealed that he had been nominated a fellow of the Geological Society: "When you pay London a visit I shall be very proud to take you [there]," he wrote.[78] And on May 16, 1838, he wrote to his sister: "I stayed at Henslow's house . . . [in Cambridge]. My friends gave me a most cordial welcome. Indeed, I was quite a lion there."[79]

Darwin was also pleased with the success of some of his early papers on geology. "I have read some short papers to the Geological Society," he wrote Fox in July 1837, "and they were favorably received by the great guns, and this gives me much confidence, and I hope not a very great deal of vanity, though I confess I feel too often like a peacock admiring his tail."[80]

How ambitious was Darwin? This question is not easy to answer. One thing is certain: his career in geology did not seem sufficiently spectacular to satisfy him. The magnitude of his friend and mentor's contribution to geology also seems to have posed a problem for him. Lyell's presence in geology at the time was considerable. The *Principles* had set a whole new course for geology. After Lyell, there was simply not much of seminal importance for Darwin to accomplish, especially after he had overseen (by 1846) the analysis, preparation, and publication of his South American geological materials. And with the onset of ill health, there was not much of a chance for him to discover new materials.

That Darwin felt that his own ideas came half out of Lyell's brain was at once an acknowledgment of Lyell's influence and an admission of the constraint Lyell's work put on Darwin's own creative abilities and ambitions. Maybe that is what Darwin meant when he said that it would have been better, in some respects, if he had not seen things through Lyell's eyes so often.

A man of Darwin's ambition and talent probably could not remain a disciple to anyone for long. We have an early indication of Darwin's uneasy view of the role of the disciple in science. Sometime in 1837 or 1838—the exact date is not known—Darwin hurriedly scribbled some pencil notes on scraps of paper. The subject of his scribblings was his future—whether or not he should marry, whether he should travel, how he might earn a living, what subjects he might study. Toward the end of his notes, he wrote: "I have so much more pleasure in direct observation, that I could not go on as Lyell does, correcting and adding up new information to old train, and I do not see what line can be followed by man tied down to London. In country—experiment and observation on lower animals,—more space—."[81]

Here Darwin seems to be choosing between subservience to Lyell and Uniformitarianism—adding new geological information to an already-postulated theory, which Darwin had done and Lyell would continue to do—and breaking out on his own, away from London, with original work on species ("experiment and observation on lower animals"). In his musings, he prefigured his later choice of the second course.

A New Friendship

By 1846, continually ill, his career in geology at a standstill, Darwin was embarking on a detailed zoological study which would last years beyond his wildest expectations and whose value he would later seriously question. Yet his situation was not without promise. He had rejected the role of disciple to Lyell, and he had met and befriended a person who would be the single most important influence in the final development of his ideas on species. This new friend was the young botanist, Joseph Dalton Hooker.

4

Working at Species

Darwin and Hooker

Joseph Dalton Hooker

Charles Lyell was Charles Darwin's senior by twelve years. Darwin looked to him, as we have seen, for leadership and advice. He was someone to emulate. Joseph Dalton Hooker was eight years Darwin's junior. Darwin looked to him for information and criticism. To Hooker, Darwin was someone to emulate. Lyell was bold, imaginative, aggressive, descisive; he was one of the "lions" of scientific London. Hooker was conservative, systematic, reserved, at times uncertain; the son of a prominent naturalist (the father, William Hooker, later became director of the Royal Botanical Gardens at Kew), he was just beginning his scientific career when he became Darwin's friend. While Lyell later would prod Darwin into action, push him toward publication, Hooker would do just the opposite: slow Darwin down with piercing criticism, ask tough questions, demand evidence and proof. Hooker, by serving as Darwin's personal sounding board, helped give shape to Darwin's ideas on species in the period 1843–59.[1]

Hooker first met Darwin in London in 1839. Their first meeting, in Trafalgar Square, was casual. "I was walking with an officer who had been . . . [Darwin's] shipmate for a short time in the *Beagle* seven years before, but who had not, I believe, since met him . . . ," Hooker later remembered. "I was introduced; the interview was of course brief, and the memory of him that I carried away and still retain was that of a rather tall and rather broad-shouldered man, with a slight stoop, an agreeable and animated expression when talking, beetle brows, and a hollow but mellow voice; and that his greeting of his old acquaintance was sailor-like —that is, delightfully frank and cordial."[2]

In the spring of 1839, the proofs of Darwin's *Journal of Researches* had reached Hooker through Charles Lyell of Kinnordy (Charles Lyell's father), who was a friend of the Hooker family. This was at a time when Hooker was hurrying to finish his medical degree at the University of Glasgow before sailing as assistant surgeon (though actually as naturalist) with Sir James Clark Ross's Antarctic expedition aboard *H.M.S. Erebus*.[3] Pressed for time, Hooker slept with the proofs under his pillow, reading them before getting out of bed in the morning. "They impressed me profoundly," Hooker later remembered; "I might say despairingly, with the variety of acquirements, mental and physical, required in a naturalist who should follow in Darwin's footsteps, whilst they stimulated me to enthusiasm in the desire to travel and observe. . . . It has been a permanent source of happiness to me that I knew so much of Mr. Darwin's scientific work so many years before that intimacy began which ripened into feelings as near to those of reverence for his life, works, and character as is reasonable and proper."[4] On the eve of Hooker's departure, the elder Lyell sent him a published copy of Darwin's *Journal*.

The *Erebus* sailed from England on September 29, 1839. In addition to magnetic survey work, its mission was the collection of various objects of natural history. There were three winter breaks during the expedition—in Tasmania, New Zealand, and the Falkland Islands. During these breaks Hooker had ample opportunity for botanical collecting. He had derived from his father an intense passion for botanical research—a passion which the *Erebus* voyage allowed him to indulge freely.

Upon his return to England in December 1843, Hooker began to prepare for publication the botanical results of the voyage. Aided with a £1000 grant from the British Treasury for plates and two small (£350) honoraria from the colonies of New Zealand and Tasmania, between 1844 and 1860 Hooker published six volumes on the flora of the *Erebus* expedition: two volumes on the flora of the Antarctic Islands (1844–47), two volumes on New Zealand (1852–54), and two volumes on Tasmania (1855–60).[5]

In 1845, after unsuccessfully attempting to secure the chair of botany at Edinburgh University (though he had the support of such leading contemporary naturalists as Robert Brown and Humboldt), Hooker was appointed botanist to the Geological Survey.

Between 1845 and 1855, his work in fossil botany received widespread praise and recognition.

In 1847 Hooker was elected a Fellow of the Royal Society[6] and in the same year received a £400 Government grant to explore the Central and Eastern Himalayan areas. In 1848–49 he examined Sikkim and eastern Nepal. He single-handedly explored the passes into Tibet. His observations on the geology and meteorology of Tibet were considered noteworthy. In 1850 he joined the British naturalist Thomas Thomson on an expedition to eastern Bengal and the Khasia Hills areas of India. He returned to England in 1851.

As a result of his Himalayan and Indian expeditions, Hooker had a collection of some seven thousand species of plants. The British Government provided him £1,200 over a three-year period to put his specimens in order—to name the species, to distribute duplicates, and so on—and to prepare a report of his findings. This report was published in 1854 as Hooker's *Himalayan Journals*.[7] In 1855, he published *Illustrations of Sikkim—Himalayan Plants*. In the same year he was appointed assistant director of Kew Gardens and published with Thomson the first volume of their *Flora Indica*.[8] In 1865, Hooker's father died, and he succeeded him as director of Kew, where he remained for the next twenty years.

Enlisting Hooker's Help

There were several references to Darwin in letters Hooker sent home during the course of the *Erebus* voyage. A few of these letters were sent to Darwin by the elder Lyell. Darwin seemed impressed with Hooker's work. He wrote to Hooker's father on March 12, 1843: "When you next write to your son, will you please remember me kindly to him. . . . I had the pleasure yesterday of reading a letter from him to Mr. Lyell of Kinnordy, full of the most interesting details and descriptions, and written (if I may be permitted to make such criticism) in a particularly agreeable style."[9]

Soon after Hooker's return to England in 1843, a lively correspondence developed between the two men. Darwin initiated contact with Hooker (as Lyell had done several years earlier with Darwin), congratulating him on completion of the voyage, suggesting some lines of investigation Hooker might wish to follow, and offering him his collection of plants from the Galapagos Islands, Patagonia, and Fuegia for examination. These collections were in the possession of Henslow at Cambridge. As Darwin wrote in December 1843:

My Dear Sir,—I had hoped before this time to have had the pleasure of seeing you and congratulating you on your safe return from your long and glorious voyage. But as I seldom go to London, we may not yet meet for some time—without you are led to attend the Geological Meetings.
  I am anxious to know what you intend doing with all your materials—I had so much pleasure in reading parts of some of your letters, that I shall be very sorry if I, as one of the public, have no opportunity of reading a good deal more. . . . I have long thought that some general sketch of the Flora of the point of land, stretching so far into the southern seas, would be very curious. Do make comparative remarks on the species allied to the European species, for the advantage of botanical ignoramuses like myself. . . . Do point out in any sketch you draw up, what genera are American and what European, and how great the differences of the species are, when the genera are European, for the sake of the ignoramuses.
  I hope Henslow will send you my Galapagos plants. . . . A flora of this archipelago would, I suspect, offer a nearly parallel case to that of St. Helena, which has so long excited interest. . . .[10]

Darwin apparently saw an opportunity to enlist Hooker's help in his researches on the geographical distribution of organisms in nature, a subject with which Darwin was intimately concerned at the time. Hooker could provide Darwin with extensive information regarding plant distribution to complement Darwin's own researches on the geographical distribution of animals.

"This [Darwin's letter] led to me sending him an outline of the conclusions I had formed regarding the distribution of plants in the southern regions," Hooker later recalled, "and the necessity of assuming the destruction of considerable areas of land to account for the relations of the flora of the so-called Antarctic Islands. I do not suppose that any of these ideas were new to him, but they led to an animated and lengthy correspondence full of instruction."[11]

Soon thereafter Darwin invited Hooker to a breakfast at his brother Erasmus's home in Park Street, London. Their friendship began instantly and flourished. "What a good thing is community of tastes! I feel as if I had known you for fifty years," Darwin revealed to Hooker in 1845.[12] And he was already touting his new friend to his now old friend Lyell: "Young Hooker talks of coming; I wish he might meet you,—he appears to me a most engaging young man."[13]

In their correspondence, the formal "My dear Sir" of the period gave way in early 1844 to "Dear Hooker" and "Dear Darwin," while the proper "very truly" or "very sincerely" were superseded by the informal "ever yours." "I hope you will excuse the freedom of my address," Darwin wrote to Hooker on February 23, 1844, "but I feel that as co–circum–wanderers and as fellow labourers (though myself a very weak one) we may throw aside some of the old world formality. . . . "[14] Eventually, both men used the even more informal "Your affectionate friend" or "Yours affectionately".

A Very Presumptuous Work

A good indication of the speed with which their friendship developed is that less than a month after first exchanging letters with Hooker, Darwin revealed to Hooker his new species hypothesis. Hooker was the first person to be thus informed. "Besides a general interest about the southern lands, I have been now ever since my return engaged in a very presumptuous work, and I know no one individual who would not say a very foolish one . . . ," Darwin cautiously wrote on January 11, 1844:

I was so struck with the distribution of the Galapagos organisms, &c. &c., and with the character of the American fossil mammifers, &c. &c., that I determined to collect blindly every sort of fact, which could bear any way on what are species. I have read heaps of agricultural and horticultural books, and have never ceased collecting facts. At last gleams of light have come, and I am almost convinced (quite contrary to the opinion I started with) that species are not (it is like confessing a murder) immutable. Heaven forfend me from Lamarck nonsense of a "tendency to progression," "adaptations from the slow willing of animals," &c.! But the conclusions I am led to are not widely different from his; though the means of change are wholly so. I think I have found out (here's presumption!) the simple way by which species become exquisitely adapted to various ends. You will now groan, and think to yourself, "on what a man have I been wasting my time and writing to." I should five years ago, have thought so. . . .[15]

For the next fifteen years, Darwin worked at his species theory, and Hooker provided constant aid and assistance. A voluminous correspondence between the two on matters directly pertaining or closely related to the species question attests to the extensive support Darwin sought and the continuing help Hooker seemed glad to provide.

It might be interesting to speculate why Darwin chose Hooker for his confidant and special assistant in this effort. Hooker was certainly young and eager to help. As Darwin's junior, he could work for Darwin in a way that someone older and more prominent, such as Charles Lyell, could not. And although young, Hooker had considerable experience in an area that Darwin knew almost nothing about—botany.[16] Hooker also had ready access to the botanical collections of the Linnean Society in London and after 1841, when Hooker's father became director of Kew Gardens, to the resources of that institution as well.

More importantly, the subject of geographical distribution, which so interested Darwin, was a topic of paramount interest to young Hooker as well. He shared with Darwin the opinion that the geographical distribution of plants and animals might be a key to understanding the origin of species, because to the trained eye the distribution of organisms can often be a clue to their development. "I know I shall live to see you the first authority in Europe on that grand subject, that almost keystone of the laws of creation, Geographical Distribution," Darwin wrote to Hooker in 1845.[17] And in the preface to his *Flora Antarctica*, Hooker described geographical distribution as the "key which will unlock the mystery of the species."[18] One must remember that the geographical distribution of animals on the Galapagos Islands seems to have first set Darwin thinking about the possibility of the mutability of species. This is why, as Darwin mentioned in his December 1843 letter to Hooker, he wanted Hooker to look at his collection of plants from these islands; he wanted to know whether the botany of this isolated group of islands was as suggestive as the zoology.[19]

Also, with regard to geographical distribution, Hooker was as aware as Darwin of the part played by geologic change in the problems of distribution. Both men were concerned with the question of how different forms of life had reached their present habitats, though Hooker was more concerned with how they were distributed and Darwin with how they originated.

Finally, Darwin had found that when he poured out his various schemes of research and his ideas for experiments to Hooker, he could always expect, if not acceptance, at least sympathy. It was "a pleasure" for Darwin to write to Hooker, Darwin would say; he had "no one to talk to about such matter" as he and Hooker talked;[20] Hooker was the "one living soul from whom . . . [I] constantly received sympathy . . . ." Darwin vowed that he would "never forget for even a minute how much assistance" he had received from him.[21]

Helping Darwin

Hooker, as special aide and extraordinary worker, performed a multitude of tasks for Darwin. His primary role was to answer Darwin's innumerable questions, almost all relating to species: "Did you collect sea-shells in Kerguelen-land? I should like to know their character."[22] ". . . My question is whether there is any relation between the ranges of genera and of individual species, without any relation to the size of the genera. . . ."[23] "Have you any good evidence for absence of insects in small islands?"[24] "Can you think of cases in any one species in genus, or genus in family, with certain parts extra developed, and some adjoining parts reduced?"[25]

Closely related to the questions were numerous observations and investigations that Darwin asked Hooker to undertake. We have already seen Darwin direct Hooker's attention to the importance of correlating the Fuegian flora with that of the Cordillera and of Europe, and invite him to study the botanical specimens that Darwin had brought back from the Galapagos Islands, as well as his Patagonian and Fuegian plants. There were numerous other such assignments.

In a letter written to Hooker January 11, 1844, Darwin proposed that Hooker "observe . . . whether any species or plant, peculiar to any island, as Galapagos, St. Helena, or New Zealand, where there are no large quadrupeds, have hooked seeds. . . ."[26] And in a November 1844 letter, Darwin suggested what "a curious, wonderful case is that of the *Lycopodium*" for studying variation, and added, "I trust you will work the case out, and, even if unsupported, publish it, for you can surely do this with due caution."[27]

In the spring of 1847, Hooker was about to leave for Nepal and

India. Darwin wrote to him that although Hooker would soon be far away, he still planned to "get some work out of . . . [him], about the domestic races of animals in India. . . . "[28] And in a May 10, 1848 letter to Hooker, then in Calcutta, Darwin, after thanking Hooker for providing him with "facts for my Species-book," offered further guidance: "Do not forget to make inquiries about the origin, even if only traditionally known, of any varieties of domestic quadrupeds, birds, silkworms, etc. Are there domestic bees? If so, hives ought to be brought home."[29]

This direction is reminiscent of the guidance Henslow gave Darwin during the *Beagle* voyage. In terms of Hooker's already completed Antarctic travels, Darwin could hope to instruct with regard only to what had already been collected. But in all of Hooker's further travels, Darwin, now something of a "lion" in his own right, could attempt to guide the collecting process itself.

An additional duty for Hooker concerned Darwin's experiments, a subject discussed in more detail below. As assistant director and later director of Kew Gardens, Hooker had access to a team of botanists and to the latest botanical research facilities. Some of Darwin's experiments were repeated under Hooker's direction at Kew; while others were first attempted at Kew in response to Darwin's request. ". . . Would not this be a curious and valuable experiment," Darwin asked in an 1845 letter to Hooker, "viz., to get seeds of some alpine plant, a little more hairy, etc., etc., than its lowland fellow, and raise seedlings at Kew: if this has not been done, could you not get it done? Have you anybody in Scotland from whom you could get the seeds?"[30] And in an 1855 letter Darwin suggested: "If you have several of the Loffoden seeds, do soak some in tepid water, and get planted with the utmost care: this is an experiment after my own heart. . . . "[31]

Still another area where Hooker gave Darwin help was in the review of Darwin's numerous manuscripts and proof sheets of his articles and books. Hooker's critical faculties and considerable knowledge of geographical distribution were of particular aid here. "And now I am going to beg almost as great a favour as a man can beg of another," Darwin wrote to Hooker on July 13, 1856, "and I ask some five or six weeks before I want the favour done, that it may appear less horrid. It is to read, but well copied out, my pages (about forty!!) on Alpine floras and faunas, Arctic and Antarctic floras and faunas, and the supposed cold mundane period. It would be really an enormous advantage to me, as I am sure otherwise to make botanical blunders."[32] And in a letter dated March 2, 1859, Darwin, after finishing a chapter on geographical distribution, asked Hooker to read it if he was not "extra busy." "On my honour, I will not be mortified, and I earnestly beg you not to . . . [read] it, if it will bother you. I want it, because I here feel especially unsafe, and errors may have crept in."[33]

Another duty of Hooker's was to provide Darwin with various articles and books which he needed for his species work and was unable to secure. Darwin was relatively isolated at Down, while Hooker, in London, had ready access to major library facilities. In a letter of April 18, 1847, Darwin asked for a copy of an English natural history journal: ". . . could you sometime spare it? I would go through it quickly. . . ."[34] And in a letter dated July 14, 1857, Darwin requested "most earnestly" a favor: "viz., the loan of *Boreau, Flore du centre de la France, either 1st or 2nd edition*, last best; also 'Flora Ratisbonensis,' by Dr. Fürnrohr. . . . If you can *possibly* spare them, will you send them at once to the enclosed address."[35]

In addition to the tasks and duties described above, Hooker provided Darwin with a whole range of miscellaneous "support services." These services were performed either in response to a request or at Hooker's own initiative. They included supplying Darwin statistical information, and often tabulating the statistics; summarizing the results of his own researches, often long before he would ordinarily do so, at great inconvenience and loss of time alerting Darwin to various books and journal articles that might be of help in his various researches; securing lending privileges for Darwin at the Library of the Linnean Society; lending Darwin books from his own library, often allowing him to make pencil annotations, and sometimes sending specific books to Darwin time and time again as his researches warranted; allowing Darwin to "pump" him (as Darwin phrased it)[36] for information when visiting at Down; and often staying for weeks at Down (bringing his own work with him) when ill health made any sort of movement for Darwin difficult.

Hooker's selfless assistance became so valued and so much a part of his everyday research activities that Darwin came to view him as an almost indispensable aid in his work.

Thinking that Hooker might leave England for good to assume

the Chair of Botany at Edinburgh University, Darwin wrote despondently to him on February 10, 1845: "I had looked forward to [our] seeing much of each other during our lives. It is a heavy disappointment; and in a mere selfish point of view, as aiding me in my work, your loss is indeed irreparable. . . . Indeed, I cannot pity you much, though I pity myself exceedingly in your loss."[37] And when in 1847 Hooker was preparing to leave for the Himalayas and India, Darwin sent him congratulations, but then added the bittersweet sentence: "It will be a noble voyage and journey, but I wish it was over. I shall miss you selfishly and all ways to a dreadful extent."[38]

Even short absences by Hooker troubled Darwin. In a letter to Hooker dated April 7, 1847, Darwin revealed his concern over a month-long holiday by Hooker. "I was much disappointed at missing my trip to Kew," he wrote, "and the more so, as I had forgotten you would be away all this month. . . . I shall feel quite lost without you to discuss many points with, and to point out (ill luck to you) difficulties and objections to my species hypotheses."[39]

Darwin in fact used Hooker so much in support of his own work that he came to feel that he had perhaps imposed too greatly on his friend's generosity. He feared that his questions and assignments had diverted Hooker from his own important researches. "I congratulate myself in a most unfair advantage of you," he wrote to Hooker in 1845, "viz., in having extracted more facts and views from you than from any one other person";[40] in a letter of April 19, 1855, Darwin admitted: "In truth, I fear I impose far more on your great kindness, my dear Hooker, than I have any claim."[41]

In the fall of 1854, having concluded his Cirripedes research, Darwin was about to begin to look over his already large collection of notes on species. The fact that he would then be concentrating full-time on species work meant that he would be relying on Hooker even more than usual. Darwin was concerned lest he, as he phrased it in a September letter to Hooker, "'progress' into one of the greatest bores in life, to the few like you with lots of knowledge."[42]

Darwin frequently warned Hooker that if he felt that Darwin was in any way imposing, he should mention the fact and all requests would cease. ". . . I fear that you answer me when busy and without inclination (and I am sure I should have none if I was as busy as you)," Darwin wrote to Hooker on May 6, 1847,

". . . pray do not do so, and if I thought my writing entailed an answer from you . . . it would destroy all my pleasure in writing."[43] And in a letter dated June 10, 1855, Darwin noted somewhat anxiously: "I hope my letters do not bother you. Again, and for the last time, I say that I should be extremely vexed if ever you write to me against the grain or when tired."[44]

Darwin seemed especially disturbed by Hooker's long letters to him. Darwin knew the great amount of time such letters took, and, although thankful for them, he could not conceal his feelings of guilt and shame: "I am almost grieved, when I saw the length of your letter, that you should have given up so much time to me," he wrote in 1844;[45] "I am going to ask you some questions, but I should really sometimes almost be glad if you did not answer me for a long time, or not at all, for in honest truth I am often ashamed at, and marvel at, your kindness in writing such long letters to me," he wrote in 1845 or 1846;[46] and in a letter on May 10, 1848, Darwin complained: ". . . I felt almost sorry when I beheld how long a letter you had written. I know that you are indomitable in work, but remember how precious your time is, and do not waste it on your friends, however much pleasure you may give them. Such a letter would have cost me a half-a-day's work."[47]

Also disturbing to Darwin was the fact that while Hooker was working hard to aid his researches, Darwin was himself doing very little to support Hooker in return. "I wish I ever had any books to lend you in return for the many you have lent me . . . ," he wrote in September 1845;[48] while in a letter sent at the end of February 1846, he noted: "If you come to any more conclusions about polymorphism, I should be very glad to hear the result: It is delightful to have many facts fermenting in one's brain, and your letters and conclusions always give one plenty of this same fermentation. I wish I could even make any return for all your facts, views, and suggestions."[49]

One incident in this regard was particularly upsetting. In 1859, Darwin had declined Hooker's request to review the proof sheets of the *Introduction to Tasmanian Flora*. This was quite surprising and unusual, as Hooker had never refused to review various of Darwin's works in manuscript or in proof form. In a letter to Hooker dated September 11, 1859, Darwin tried to explain and apologize: "I write now to say that I am uneasy in my conscience about hesitating to look over your proofs, but I was feeling miserably unwell and shattered when I wrote. I do not suppose I could be of hardly any use, but if I could, pray send me any proofs. I should be (and fear I was) the most ungrateful man to hesitate to do anything for you after some fifteen or more years' help from you. . . . I hope to God, you do not think me a brute about your proof-sheets."[50]

A Magnitude of Topics

Hooker was nevertheless undaunted in his support. As a good soldier or devoted companion might, he responded to Darwin's every call. One senses the magnitude of the response from the large range of topics they discussed: physiological compensation, the effects of climate, the variability of large genera, botanical analogues to zoological development, hybrids, multiple creation, Cirripedes, the question of organic "highness" or "lowness," disuse, reversion, affinities, polymorphism, land mollusca, domestic breeding, population pressure and struggle, the variability of abnormal organs, extinction, representative species, isolation, the range of mundane species.

Some of the topics were hotly, though lightheartedly, debated, as in the case of coal. The question concerning coal centered on how that substance had been formed. Darwin leaned heavily toward the theory that coal developed originally in seashore, swamplike areas as a sort of submarine peat. Hooker was not sure, though he thought Darwin's idea of the submarine origin of coal absurd. The two were often at each other's throats on the issue. "You have made a savage onslaught [against my coal theory], and I must try to defend myself . . . , " Darwin wrote half-jokingly on May 6, 1847.[51] And in a letter later that year to Hooker, Darwin humorously revealed: ". . . as submarine coal made you so wrath, I thought I would experimentise on Falconer and Bunbury . . . and it made [them] even more savage. . . . So I now know how to stir up and show off any Botanist."[52]

Another, more important discussion, which at times also turned into heated though convivial debate, centered on the question of the geographical distribution of organisms and the means of their dispersal. The discussion focused on the question of continental extension.

In 1846, the English naturalist Edward Forbes[53] had suggested that at one time the major continents had been connected to each other by large land bridges, now subsided. He went so far as to suggest the possible existence of a lost Atlantic continent, "Atlantis." Forbes's theory applied as well to the connecting links between continents and major oceanic islands. Darwin was dismayed that both Hooker and Lyell, in their attempts to explain the geographical distribution of plants and animals, were carrying some of Forbes's ideas to a point where they were creating new continents, as Darwin phrased it, "as easily as a cook does pancakes."[54]

The arguments on both sides quickly heated up. To Lyell, on June 16, 1856, Darwin facetiously complained: "I am going to do the most impudent thing in the world. But my blood gets hot with passion and turns cold alternately at the geological strides, which many of your disciples are taking. Here, poor Forbes made a continent to [i.e., extending to] North America and another (or the same) to the Gulf weed; Hooker makes one from New Zealand to South America and round the World to Kerguelen Land. Here is Wollaston speaking of Madeira and P. Santo 'as the sure and certain witnesses of a former continent.' Here is Woodward writes to me, if you grant a continent over 200 or 300 miles of ocean depths (as if that was nothing), why not extend a continent to every island in the Pacific and Atlantic Oceans? And all this within the existence of recent species! If you do not stop this, if there be a lower region for the punishment of geologists, I believe, my great master, you will go there. Why, your disciples in a slow and creeping manner beat all the old Catastrophists who ever lived. You will live to be the great chief of the Catastrophists."[55]

To Hooker, who, though also not a blind follower of Forbes, was still too liberal for Darwin on the question, Darwin wrote on July 30, 1856: "There never was such a predicament as mine: here you continental extensionists would remove enormous difficulties opposed to me [in my species theory], and yet I cannot honestly admit the doctrine, and must therefore say so . . . ";[56] a few days later, on August 5, Darwin wrote again: "You cannot imagine how earnestly I wish I could swallow continental extension, but I cannot; the more I think (and I cannot get the subject out of my head), the more difficult I find it. . . . I cannot but think that the theory of continental extension does do some little harm as stopping investigation of the means of dispersal. . . . "[57]

Darwin's objections to Forbes's theory were these: first, the coasts of America and Africa contain shells that have been distinct since the Miocene period. This indicates that these coasts have remained substantially as they are today for a length of time infinitely longer than the life-span of existing species. Second, almost all oceanic islands (except St. Paul's Rocks and the Seychelles), because of volcanic or coralline origin, do not require continental extensions to explain their existence. Because of their origin, almost all oceanic islands present a picture totally different from islands which would result from the deep submergence of existing continents. Finally, if indeed continental extensions and land bridges had existed, how could one explain why so many plants and animals today, particularly amphibia and mammals, had not reached and populated oceanic islands and invaded other continents? To invoke a land bridge to explain the geographical distribution of five percent of species leaves one with the problem of explaining the geographical distribution of the remaining 95 percent.[58]

Instead of hypothetical continents or land bridges to account for the geographical distribution of plants and animals, Darwin felt that a more fruitful line of investigation would be to study the methods of fortuitous dispersal of organisms. Darwin had previously been working on this problem. He was eager to determine if seeds could be transported across large areas of ocean and still retain their germinating powers. He hypothesized that seeds could float across large ocean areas, or could be blown across, or might be eaten by fish or birds that fly long distances and then deposited through vomiting or as part of their droppings. Darwin also thought that seeds might be transported in the caked dirt of birds' feet or in their crops. Seeds could also have been carried by icebergs.

Darwin conducted a series of experiments to test these hypotheses. The existing popular belief was that seeds were instantly destroyed by seawater. Darwin made saltwater tanks and tested the power of seeds to sink or swim. Most experiments were performed on temperate and tropical seeds. The seeds were often supplied by Hooker at Kew, where some of the experiments were also conducted. Darwin's investigations, however, were not confined to seeds, but included eggs and snails as well. In addition, Darwin was concerned not only with the effects of immersion in water but also with desiccation in the air. Finally, Darwin carefully studied the speeds and directions of ocean currents flowing past the shores of continents toward oceanic islands.

The experiments became a point of good-humored repartee between Darwin and Hooker, and there was keen anticipation concerning their results. Hooker, the conservative doomsayer, scoffed and doubted, while Darwin, perhaps naively, moved credulously ahead.

"When I wrote last I was going to triumph over you, for my experiment had in a slight degree suceeded . . .," Darwin wrote to Hooker in a letter dated April 13, 1855, "but this, with infinite baseness, I did not tell, in hopes that you would say that you would eat all the plants which I could raise after immersion. It is very aggravating that I cannot in the least remember what you did formerly say that made me think you scoffed at the experiments vastly; for you now seem to view the experiment like a good Christian."[59]

The following day, Darwin wrote again: "You are a good man to confess that you expected the crest would be killed in a week, for this gives me a nice little triumph. The children at first were tremendously eager, and asked me often, 'whether I should beat Dr. Hooker!' "[60] As successes began to mount up, Darwin's confidence increased dramatically, to the point where he kiddingly boasted to Hooker: "I believe you are afraid to send me a ripe Edwardsia pod for fear I should float it from New Zealand to Chile!"[61]

But Hooker was still not convinced. He finally planted several seeds himself that the Gulf Stream had carried across the Atlantic to the coast of Norway to see if they would germinate. They germinated perfectly and to Hooker's confession of defeat, Darwin responded on June 1, 1856: "I read your note as far as 'unutterable mortification' and was in despair, for I came instantly to the conclusion that probably Government had determined to give up Kew Gardens! and you may imagine how I laughed when I came to the real cause of mortification. It is the funniest thing in the world that you do not rejoice; for you have (as I never have) put in print that you do not believe in multiple creation, and therefore you surely should rejoice at every conceivable means of dispersal. Well, I and my wife have enjoyed a jolly laugh, and all the more from fully believing for a second that some great calamity had befallen you."[62]

Some of the results of the experiments were interesting and

seemed to substantiate conclusively Darwin's point that continents and oceanic islands could indeed have been populated through a process of fortuitous dispersal. Of 87 species of seeds immersed for 28 days, 64 germinated. Nuts germinated after 90 days' immersion, and some seeds after 137 days. If the average rate of flow of Atlantic Ocean currents is 33 miles a day, even the survivors of only 28 days' immersion might have traveled nearly a thousand miles![63]

Arguing Species

From our point of view, however, the most important topic that Darwin and Hooker discussed was Darwin's theory of the evolution of species. It is one of the most interesting aspects of their relationship that despite their close friendship, despite the enormous help Hooker provided Darwin, and despite Hooker's close involvement with Darwin's species theory, Hooker remained, almost up until the publication of the *Origin*, skeptical of Darwin's ideas on the subject. He was always sympathetic, but never really enthusiastic and quite often severely critical.

"Bother variation, development and all such subjects!" Hooker wrote to Darwin in July 1845; "it is reasoning in a circle I believe after all. As a Botanist I must be content to take species as they *appear to be*, not as *they are*, and still less as they were or ought to be."[64] "I am amused at your anathemas against variation and co.;" Darwin replied, "whatever you may be pleased to say, you will never be content with simple species, 'as they are.' "[65]

For the longest period, however, as the correspondence between the two men reveals, Hooker in fact was content. This made Darwin apprehensive and uneasy. "I am very glad to hear that you mean to attack this subject [of evolution] some day, I wonder whether we shall ever be public combatants . . . ," Darwin wrote in 1845.[66] And in a letter of April 18, 1847 to Hooker, Darwin complained: "I see you have introduced several sentences against us Transmutationists."[67] At one point, May 10, 1848, Darwin wrote that even though Hooker might "wish my barnacles and species theory al Diavolo together," Darwin did not care. ". . . My species theory is all gospel!" he facetiously declared.[68]

In 1854 Darwin had cause for concern. The distinguished American naturalist Asa Gray seemed to be lining up with Hooker against Darwin. "I am particularly obliged to you for sending me Asa Gray's letter . . . ," he wrote to Hooker on March 26 of that year, "to see his and your caution on the species-question ought to overwhelm me in confusion and shame; it does make me feel deuced uncomfortable."[69]

Gray's and Hooker's caution did not, however, deter Darwin for long, though he continued to view Hooker for some time in a possibly adversarial role. "In a year or two's time, when I shall be at my species book (if I do not break down), I shall gnash my teeth and abuse you for having put so many facts so confound-edly well," Darwin wrote in 1854;[70] while in a letter of July 13, 1856, Darwin noted with disappointment: "I have been working your books as the richest (and vilest) mine against me. . . ."[71]

By 1858, Darwin had muted many of Hooker's sharpest criti-cisms. "I am now working several of the large local Floras, with leaving out altogether all the smallest genera . . . ," Darwin wrote to Hooker on February 28: "I shall then show how my theory points, how the facts stand, then state the nature of your grievous assault and yield entirely or defend the case as far as I can hon-estly. . . . " Darwin then added: "I have not felt the blow [of your criticisms] so much of late. . . . "[72]

Hooker felt that species were more variable and less easily defined than most naturalists then believed. He accepted this var-iability and the spread of species through migration; he even accepted their relationship to allied species and the existence of fossil predecessors in the same and closely related areas. Yet he would not accept Darwin's species theory.

Darwin seemed so concerned about the implications of Hook-er's opposition that in late 1858 he begged his friend "not to pro-nounce too strongly against Natural Selection, till you have read my abstract, for though I daresay you will strike out *many* diffi-culties, which have never occurred to me; yet you cannot have thought so fully on the subject as I have."[73] Yet Hooker's skepti-cism persisted. On December 26, 1858, he wrote to Darwin: "Your letter has interested me more than any you every wrote me (be-cause we are both *ripening I hope*), but it staggers me too. It opens a much wider question upon which I have often pondered in vain and have hoped latterly to have made more of: it is this—are we right in assuming that the development of plants has been parallel to that of animals? I sent out a feeler in the concluding notices of my review of A. De Candolle where I indicate my view that Geology gives no evidence of a progression in plants."[74]

Even as late as March 2, 1859, Darwin was inquiring of Hooker which parts of his chapter on geographical distribution Hooker would "*most vehemently* object to";[75] and on May 11, 1859, he wrote to Hooker: "Thank you for telling me about obscurity of style. But on my life no nigger with lash over him could have worked harder at clearness than I have done. . . . I imagine from some expressions (but if you ask me what, I could not answer) that you look at variability as some necessary contingency with organisms, and further that there is some necessary tendency in the variability to go on diverging in character or degree. *If you do*, I do not agree . . . it was on such points as these *I fancied* that we perhaps started differently."[76]

Two Very Different People

Why did Hooker have so much difficulty accepting Darwin's ideas? One reason was simply that Darwin gave him the role of reviewer and critic, which seemed to indicate that he was expected to criticize Darwin's works, his species work included. By being critical, Hooker was living up to his expected function, as he seemed determined to do.

Also, Hooker had difficulty understanding Darwin's species ideas, and Darwin had trouble clearly explaining them. For an understanding of Darwin's view of his own theory and its presentability as well as later reactions to his ideas, it is noteworthy that even Hooker, who worked for fifteen years on a daily basis with Darwin and dealt with almost all aspects of his friend's theory, had difficulty understanding Darwin's ideas.

Referring to the species theory, Hooker wrote to Darwin in July 1845: "I am annoyed at my own incapacity to fathom or follow the subject to any good purpose (open confession is *good* for the soul)."[77] And in a letter to Hooker dated August 5, 1856, Darwin noted: "I see from your remarks that you do not understand my notions (whether or no worth anything) about modification; I attribute very little to the direct action of climate, &c."[78] As late as January 20, 1859, Hooker was apparently still having problems. "When you say you cannot master the train of thoughts," Darwin wrote to him on that day, "I know well enough that they are too doubtful and obscure to be mastered. I have often experienced what you call the humiliating feeling of getting more and more involved in doubt the more one thinks of the facts and reasoning on doubtful points."[79]

Darwin took most of the blame himself for Hooker's inability to understand. In a March 16, 1858 letter to Hooker, Darwin felt that he had not sufficiently explained his ideas, and he asked his friend to await more detailed explanations that would become available in various chapters of the *Origin* that Darwin was then preparing.[80]

Yet another reason for Hooker's hesitancy relates to the fact that for all their friendship and closeness, Hooker and Darwin were really very different people and very different sorts of naturalists. Hooker was conservative, a systematist more than a generalizer; he was a man not eager to generalize himself. Darwin, in contrast, was more a generalizer than systematist; although in some sense as cautious as Hooker concerning the need for solid supporting evidence, Darwin was also a bold theorizer, a man who really delighted in generalization.

Hooker felt that true views of geographical distribution, for example, were impossible without full and accurate floras. To produce complete floras, Hooker needed to examine his vast materials, to reexamine what he saw as the incomplete work of predecessors, to sweep away existing redundancy and overlapping, to readjust systematic details to make clear the true affinities and geographical distribution of disputable genera and species, and to produce a complete and accurate classification according to nature. He approached the species question in a similar, tough-minded way. He went to India possessed by, but certainly not converted to, Darwin's ideas. In the Sikkim region of Nepal, where tropical and arctic flora met, Hooker expected to find transitional forms—the solid evidence he was seeking. But he did not. Thereafter, Hooker continued to work along the accepted lines of the fixity of species.

Hooker's cautious nature is a prominent theme in the correspondence between Darwin and himself. In his first letter to Darwin, in December 1843, Hooker admitted "not being a good arranger of extended views" and as such he was concerned about Darwin's first assignment regarding the geographical distribution of plants in the Southern regions: " . . . I rather fear the Geographical distribution, which I shall not attempt until I have worked out all the species. . . . "[81]

And in a letter to Darwin in April 1845, he commented pejoratively on the speculative tendencies of contemporary botanists: "Except Brown and Humboldt . . . all seem to dread the making Bot. Geog. too exact a science; they find it far easier to speculate than to employ the inductive process."[82]

Emphasizing the theme of the importance of firsthand direct observation, Hooker wrote to Darwin in July 1845: "I am not inclined to take much for granted from any one who treats the subject in his [the French naturalist Godron's] way and who does not know what it is to be a specific Naturalist himself. Those who have had most species pass under their hands, as Bentham, Brown, Linnaeus, Decaisne, and Miguel, all I believe argue for the validity of *species* in nature. . . . "[83] We have already seen that Hooker, in a July 1845 letter to Darwin, felt that Darwin's speculations on species were "reasoning in a circle" and that Hooker felt he must be content to accept species as they empirically appear to be, not as they are, much less as they ought speculatively to be.

A good illustration of Hooker's conservative, highly inductive, empirical approach to natural history can be seen in a letter he wrote to Darwin on the occasion of the publication of his *Himalayan Journals*, March 1, 1854. In this study he sought to describe the flora of an unexplored country. He saw this descriptive work as the culmination of his ambition. He wrote:

You will laugh when I tell you that, my book out, I feel past the meridian of life! But you do not know how from my earliest childhood I nourished and cherished the desire to make a creditable journey in a new country, and write such a respectable account of its natural features as should give me a niche amongst the scientific explorers of the globe I inhabit, and hand my name down as a useful contributor of original matter. A combination of most rare advantages has enabled me to gain as much of my object as contents me, for I never wished to be greatest amongst you, nor did rivalry ever enter my thoughts. No ulterior object has ever been present to me in this pursuit. My ambition is fully gratified by the satisfactory completion of my task, and I am now happy to go on jog-trot at Botany till the end of my days—downhill, in one sense, all the way. I shall never have such another object to work for, nor shall I feel the want of it. . . . As it is, the craving of thirty years is satisfied, and I now look back on life in a way I never could previously. There never was a past hitherto to me. The phantom was always in view. . . . [84]

Darwin later revealed that Hooker was one of the factors leading him to postpone his species work[85] because he chastised Darwin for caring more for his theoretical species work than for his
mundane Barnacles studies. And it was Hooker who was forever
putting a harness on Darwin's speculative tendencies. "Your last
very instructive letter shall make me very cautious on the hyper-
speculative points we have been discussing . . . ," Darwin wrote.[86]
Hooker warned Darwin to "take care not to get *entêté* with your
results."[87] Darwin called Hooker a "stern and awful judge and
skeptic,"[88] and at one point became a bit peeved at what he considered Hooker's sometimes impossible standards, in this case
relating to observation. "Shall you think me very impudent if I
tell you that I have sometimes thought that (quite independently
of the present case), you are a little too hard on bad observers;
that a remark made by a bad observer *cannot* be right . . . ," Darwin wrote; "an observer who deserves to be damned you would
utterly damn. I feel entire deference to any remark you make out
of your own head; but when in opposition to some poor devil, I
somehow involuntarily feel not quite so much, but yet much deference for your opinion. I do not know in the least whether there
is any truth in this my criticism against you, but I have often
thought I would tell you it."[89]

Darwin, on the other hand, often attempted to get Hooker to
generalize from his detailed observations. "How interesting the
flora of the Sandwich Islands appears to be," Darwin wrote to his
friend in 1849; "how I wish there were materials for you to treat
its flora as you have done the Galapagos. In the Systematic paper
I was rather disappointed in not finding general remarks on affinities, structure, &c., such as you often give in conversation, and
such as De Candolle and St. Hilaire introduced in almost all
their papers, and which make them interesting even to a non-
Botanist."[90]

One of the problems that Darwin faced in getting Hooker to
generalize more seemed a lack of confidence on Hooker's part in
his own theoretical capabilities. Darwin often attempted to bolster Hooker in this regard. As he wrote to Hooker on January 11,
1844, in reply to his December 1843 letter: "I must write to thank
you for your last letter, and to tell you how much all your views
and facts interest me. I must be allowed to put my own interpretation on what you say of 'not being a good arranger of extended
views'—which is, that you do not indulge in the loose specula-

tions so easily started by every smatterer and wandering collector."[91] To Hooker's contention that he did not have a philosophic mind, Darwin shot back strongly: "One of the greatest falsehoods ever told by implication; read your Galapagos paper and be ashamed of yourself."[92] Darwin felt that it was simply "unjust" for Hooker to think of himself as only a systematist.[93] Darwin maintained that Hooker was, at heart, a generalist and that he should not attempt to suppress his true capabilities. In a December 1857 letter to Darwin regarding Hooker's *Introduction to Tasmanian Flora*, Hooker suggested that he would confine himself "to a clear exposition of all the main features of the Flora of Australia and leave all conclusion drawing to others."[94] To this, Darwin responded: "I am very sorry to hear you do not intend to give generalisations in your Tasmanian Introduction but I do not believe you will be able to resist; what is in the spirit must come out."[95] In a letter of October 23, 1858, Darwin suggested that Hooker should not attempt to "delude" himself by saying that he intends "sticking to humdrum science." "I believe," Darwin continued, "it is just as much as if a plant were to say that, 'I have been growing all my life, and, by Jove, I will stop growing.' You cannot help yourself; you are not clever enough for that."[96]

Darwin felt that Hooker needed nothing so much "as a little vanity." "Then," Darwin wrote on March 31, 1845, "you would not talk of collecting facts for others, when, say just what you please, I am sure no one could put them to better use than yourself."[97] In the same year Darwin wrote to his friend: "I have never perceived but one fault in you, and that you have grievously, viz., modesty; you form an exception to Sydney Smith's aphorism, that merit and modesty have no other connection, except in their first letter."[98] When Hooker did show some ambition, Darwin was quick to compliment him. "I am glad you have shown a little bit of ambition about your Journal," he wrote to Hooker on March 1, 1854, "for you must know that I have often abused you for not caring more about fame."[99]

In time, however, Darwin came to realize that Hooker was not going to change greatly, and he tried to reassure him that observation, sometimes along with a little theory, was what science most benefited from. "Depend on it," he wrote to Hooker in June 1856, "my saying is a true one, viz., that a compiler is a *great* man, and an original man a commonplace man. Any fool can generalise and speculate; but, oh, my heavens! to get up *at second hand* a New Zealand Flora, that is work."[100] On December 21, 1859, Darwin wrote reassuringly: "I hate to hear you abuse your own work. I, on the contrary, so sincerely value all that you have written. It is an old and firm conviction of mine, that the Naturalists who accumulate facts and make many partial generalisations are the *real* benefactors of science."[101]

A Private Public

There was a running argument between Hooker and Darwin about who really gained more from their friendship. It is certain that in asking for his aid, Darwin also stimulated Hooker to some extent to deal in the wider implications of his observations, even though he was often reluctant to do so. Though Darwin's assignments often took Hooker away from his descriptive flora work, they might have also helped Hooker shape his own ideas and enhance his reputation. "I am almost sorry for your eternal additional labours on the Galapagos Flora," Darwin wrote to him in September of 1846; "as yet your work assuredly has not been thrown away, as many have referred to your curious geographical results on this archipelago."[102] In 1844 Darwin had asked Hooker to prepare a preliminary sketch of his Tasmanian results, well before Hooker had intended to. Darwin was eager to assure Hooker of the positive benefits that that particular time-consuming "assignment" would have: "I trust that your sketch will not have cost you ultimately loss of time," Darwin wrote, "as, judging by myself, preliminary sketches and resketches do much good. . . ."[103]

Hooker apparently needed no convincing. He expressed gratitude for Darwin's various questions and assignments, which helped turn him, as he put it, "from the drudgery of my 'professional Botany' to your 'philosophical Botany'."[104] He looked upon such work, he generously wrote to Darwin, "as the best means of keeping alive a due interest in . . . [broader botanical] subjects. I indulge vague hopes of treating them some day, but days and years fly over my head and all I do is done in correspondence to you, but for which I should soon lose sight of the whole matter."[105] Later, after Darwin's death, Hooker remembered his visits to Down and his conversations with Darwin: " . . . I at any rate always left with the feeling that I had imparted nothing and carried away more than I could stagger under."[106]

We have already seen how much Hooker aided Darwin. Darwin felt that he had "extracted more facts and views from . . . [Hooker] than from any one other person."[107] He also felt he had learned in conversations with Hooker "ten times over the number of hours reading"[108] would have given him. Even the quasigeneral statements Darwin was able to squeeze out of Hooker were useful, and eventually, I think, Darwin realized that he did not really need Hooker to generalize very much. He needed Hooker only to supply information; Darwin could do the generalizing himself.

Darwin needed someone to develop and test his species ideas against. The fact that Hooker had also expert knowledge in areas —botany, geographical distribution—that were crucial to an understanding of the theory Darwin was attempting to explain and substantiate made Hooker's help all the more valuable. As Hooker criticized and chided, Darwin restructured ideas, restated concepts, and looked for new evidence until ideas and a general scheme of argument began to fall into place.

Over the years, Darwin thanked Hooker in many ways for his assistance. The most poignant thanks, however, came several years after the *Origin* had appeared and summarize succinctly the important role as sounding board that Hooker played in Darwin's species work. Darwin wrote to him on September 13, 1864: "You have represented for many years the whole great public to me."[109]

5

Working at Species

Darwin and Others

A Correspondence Network

Although Joseph Hooker was the most significant person Darwin depended upon for his work on species, he was by no means the only one. Darwin gathered around him a group of people who, like Hooker, were put into service on behalf of Darwin's species work: to answer questions, perform various "assignments," provide needed materials. Putting the *Origin* together became as much an effort of organizing and managing people as of creating and testing new ideas. Darwin's work in the period from 1838 to 1859 represents the efforts not only of an imaginative thinker and bold theorizer, but also of a smart organizer and efficient manager—a charming, resourceful, and persuasive salesman who knew how to get the last bit of information out of those who had it, but who also had the capability to generate his own information where none existed.

There were a number of ways in which Darwin sought to support his work on species. He was a voracious reader. Reading pervaded the atmosphere of life at Down. Emma read letters, travels, histories, and novels aloud to him; he read stories aloud to the children and the newspaper to himself. He read a good deal of many specialized works in and outside his field. He read large parts of natural history textbooks. He read whole series of various journals. He was, in fact, so voracious a reader that he read nearly the whole of every issue of the British natural science periodical *Nature*, which dealt primarily with mathematics and physics, even though he admitted that he did not understand most of the articles.[1]

Darwin devised a methodical system for the review of books

and articles closely related to his species work. One shelf in his study contained books to be read; another shelf was reserved for those already read (or skimmed). When the second shelf became crowded, a day was reluctantly set aside for cataloging. As Darwin read each book, he marked important passages in the margin, where he also often added short notations. At the end of the book he made a list of the pages so marked. When it came time for cataloging, he reviewed marked passages and notations and made an abstract of the book based on these markings and notations. Using this method, Darwin went through scores of books as well as whole series of periodicals, producing literally hundreds of abstracts for his files.[2]

Also, Darwin was an avid and often ingenious experimenter. We have already mentioned Darwin's work on the oceanic transport of seeds and eggs. He also experimented on seeds and plants as part of his interest in the question of population pressure in nature. He had an experimental, two-by-three-foot weed garden at Down in which he tested the ability of various seeds to survive.[3] He also had a meadow planted with some sixteen different varieties of seeds, where he followed carefully the struggle for existence among various young plants.[4] In addition, with the help of Miss Thorley, the family governess, he conducted experiments on struggle utilizing two adjacent Down farm fields. One field had been allowed to run waste for fifteen years (though it had been cultivated in the past); the other was then in cultivation. Darwin planted seeds, periodically inspected the fields, and collected and compared plants from each.[5]

Darwin also performed experiments on the effects of use and disuse. In one of these he collected, weighed, and measured the skeletons of wild and tame ducks. He found the foot of the tame duck to be much heavier than that of its wild counterpart.[6] He also experimented on the movement of plants. One plant, the Hedysarum, was reported to close its leaves in darkness more quickly than any other. Darwin attempted, by covering it daily for half an hour or so, to see if he could teach it to close its leaves by itself.[7] Finally, as part of his work on variation under domestication, Darwin bred his own pigeons and, to further his knowledge in this area, became a member of several of the leading pigeon breeding clubs.[8]

There were several methods that Darwin employed to get help from others. In 1839 he developed and circulated a printed questionnaire entitled "Questions About the Breeding of Animals." He sent it to various animal breeders and farmers. It was an effort to understand more about variation under domestication. How successful it was in generating the information Darwin needed is not clear, but it was probably fairly successful because many years later he circulated a similar questionnaire, which he sent "to all parts of the world," when he was doing research for the *Descent of Man.* This second questionnaire attempted to gather information on the subject of native views of beauty.[9]

When naturalists were visiting Down, Darwin used the opportunity to solicit information from them. In the early years at Down guests came often, despite Darwin's ill health. Hooker, a frequent visitor, later described the atmosphere encountered there: "A more hospitable and more attractive home under every point of view could not be imagined—of Society there were most often Dr. Falconer, Edward Forbes, Professor Bell, and Mr. Waterhouse —there were long walks, romps with the children on hands and knees, music that haunts me still. Darwin's own hearty manner, hollow laugh, and thorough enjoyment of home life with friends; strolls with him all together, and interviews with us one by one in his study, to discuss questions in any branch of biological or physical knowledge that we had followed. . . ." Later, when Darwin's health became worse, Hooker was "for days and weeks the only visitor, bringing my work with me and enjoying his society as opportunity offered. It was an established rule that he every day pumped me, as he called it, for half an hour or so after breakfast in his study, when he first brought out a heap of slips with questions botanical, geographical, &c., for me to answer, and concluded by telling me of the progress he had made in his own work, asking my opinion on various points."[10]

Darwin's most important means of collecting information was an extensive correspondence network, developed over a course of years, which linked him, usually on a regular basis, with a host of helpers in Britain and throughout the world. This correspondence network became the chief vehicle by which Darwin was able to involve others in support of his species researches. It included such people as Hooker; T. H. Huxley; Charles Lyell; Leonard

Jenyns; the English geologist and mineralogist, S. P. Woodward; one-time Keeper of Zoology at the British Museum, John Edward Gray; Darwin's former *Beagle* servant, Sym Covington; J. S. Henslow; the Shetland Island naturalist, Laurence Edmondston; the British poultry expert, W. B. Tegetmeier; W. Darwin Fox; and Asa Gray, among others.[11]

Not everyone was used in the same way or for the same purposes. Darwin seemed to have a special talent for knowing how people could help him and for concentrating his requests for assistance in their strongest areas of expertise. Huxley answered questions on variability and later served as Darwin's bulldog—"my good and admirable agent for the promulgation of damnable heresies!," as Darwin called him;[12] Lyell was used for strategic advice about the publication of the *Origin*;[13] Jenyns for information about population pressure;[14] Woodward for various facts on species ranges;[15] John Edward Gray helped Darwin secure various books and other materials from the British museum;[16] Sym Covington provided Darwin with skins and other specimens related to Darwin's studies of domestic animals;[17] Henslow was asked to assist Darwin in researches on eggs and seeds;[18] Edmondston helped with Darwin's work on pigeons and was asked to provide Shetland pigeon specimens and any other information on the habits of domesticated animals of the Shetland Islands;[19] Tegetmeier helped Darwin with various questions related to the different breeds of fowls and pigeons.[20]

With the exception of Hooker, Fox and Asa Gray were the two in the network Darwin used most extensively. Fox was asked to answer questions, make observations, and collect specimens— pigeons, ducklings, hens, chickens, eggs—as part of Darwin's work on the variation of domestic species. In a letter dated March 19, 1855, Darwin asked for Fox's help in determining "at what age nestling pigeons have their tail feathers sufficiently developed to be counted." Darwin had never carefully observed a young pigeon. He wanted to see, with regard to the young of British species, "how young, and to what degree the differences . . . [among them] appear." Darwin felt he must either breed pigeons himself or buy them, and he wanted to know something of their development so if he did have to buy some, he would not be "excessively liable to be cheated and gulled," as he put it. "I know you will help me *if you can* with information about the young pigeons," he wrote to Fox. He also asked Fox for poultry specimens: "a chicken with exact age stated" would be helpful.[21]

A week later, in a letter dated March 27, 1855, Darwin asked Fox for old poultry hens—"should . . . any . . . die or become so old as *useless*, I wish you would send her to me per rail. . . ." In the same letter, he indicated to Fox that he was pleased with the help he was getting from others regarding his species work: "I have a *number* of people helping me in every way, and giving me most valuable assistance. . . ."[22]

Several months later, in a May 7, 1855 letter to Fox, Darwin asked for more specimens: "a seven-days' duckling and for one of the old birds, should one ever die a natural death." Also, if "an old wild turkey ever die, please remember me. . . ."[23]

By the summer of 1855, Darwin was deeply involved with species collecting, measuring, and dissecting. He already had puppies of bulldogs and greyhounds in salt, and he had had the young colts of a cart horse and racehorse measured.[24] On May 23, Darwin, seeking more specimens, asked Fox for a one-week-old mongrel chicken, in order to determine "whether the *young* of our domestic breeds differ as much from each other as do their parents. . . ." Deciding that he should breed his own pigeons, Darwin had fantail and pouter pigeons in what he called "a grand cage and pigeon house," and they were, he proudly told Fox, a delight to him.[25] In July, Darwin thanked Fox "for the seven days' old white Dorking [pigeon], and for the other promised ones."[26] Fox would later supply Darwin with information on cats as well.[27]

By the fall of 1856, Darwin felt that his work on pigeons had become invaluable, "enlightening me," as he wrote to Fox on October 3rd, "on many points on variation under domestication." It was also apparent that his search for specimens had become a worldwide effort: "I have just had pigeons and fouls *alive* from Gambia!" he excitedly wrote in the same letter. He also revealed that he had found the "most remarkable differences in the skeletons of rabbits" and questioned Fox concerning any odd breeds of rabbits that Fox might have kept for which he could provide any details.[28] Later, Fox was asked to help Darwin with observations on the stripes of horses. Darwin thought that such information would help him understand the color of the primeval horse and, presumably, how the domestic horse had evolved.[29]

Finally, Fox played an important role in Darwin's experiments on the oceanic transport of seeds and eggs. "You will hate the very sight of my handwriting;" he wrote to Fox on May 17, 1855, "but after this time I promise I will ask for nothing more, at least for a long time. As you live on sandy soil, have you lizards at all common?" He offered to reward the young boys in Fox's school for any lizards' eggs they might collect: "a shilling for every half-dozen, or more if rare." If snake eggs were brought in by mistake, that was all right too. Darwin's object was "to see whether such eggs will float on sea water, and whether they will keep alive thus floating for a month or two in my cellar." "I am trying experiments on transportation of all organic beings that I can;" he revealed to Fox, "and lizards are found on every island, and therefore I am very anxious to see whether their eggs stand sea water."[30]

Darwin used Asa Gray primarily to help with researches on geographical distribution—a subject in which, as we have seen, Darwin had a particularly strong interest. As Hooker was asked to provide Darwin with information concerning Antarctic, New Zealand, Indian, and Himalayan flora, Gray, a friend of Hooker's and a botanist, was used in a similar manner regarding the flora of North America. Darwin was especially pleased to have Gray's help, as he began to see a special role for botany in his species researches. Darwin felt that botany was conducted in a "much more . . . philosophical spirit than Zoology." As a result, he would not trust "any general remark in Zoology without I find that botanists concur."[31]

Darwin first met Gray at Kew Gardens in 1855. Gray was then visiting Hooker. In his first letter to Gray, dated April 25, 1855, Darwin requested information regarding American Alpine plants.[32] A short time later, in a June 8, 1855 letter, Darwin asked for additional information, this time about the proportion of European plants to the whole of American flora—"in order to speculate on means of transportal." Darwin also asked Gray to classify his work on American genera into species common to the Old World, stating numbers common to Europe and Asia, and species belonging to genera confined to America or to the New World. Such a classification, Darwin felt, would help clarify the migratory paths of various European plants. Also, Darwin was "most anxious" to have what he termed "close species" in a flora marked, "so as to compare in *different* Floras whether the same genera have 'close species,' and for other purposes too vague to enumerate."[33]

Gray was an enthusiastic and cordial correspondent and helper, even though he had severe reservations about the validity of Darwin's species theory. But Gray saw the usefulness of such general studies, even though, like Hooker, he did not envision himself as one who might undertake them. "I rejoice in furnishing facts to others to work up in their bearing on general questions," he candidly wrote to Darwin on June 30, 1855, "and feel it the more my duty to do so inasmuch as from preoccupation of mind and time and want of experience I am unable to contribute direct original investigations of the sort to the advancement of science."[34]

Gray became, along with Hooker and a few others, one of the small group of people Darwin was willing to confide in regarding the evidential bases for his species theory. In a revealing letter on July 20, 1856 to Gray, Darwin told how he was seeking to substantiate his ideas. Darwin's method, which seems to have been very deductive, was to assume that species were descended from others with considerable extinction and then to test this assumption "by comparison with as many general and pretty well-established propositions as I can find made out,—in geographical distribution, geological history, affinities, &c., &c." Darwin felt that if such a theory made sense in terms of general propositions in natural history, then it ought to be accepted until something better came along.[35]

Between 1855 and the publication of the *Origin* in 1859, Darwin communicated regularly with Gray, sending a steady flow of questions and assignments to his American friend. Darwin came to depend on Gray for both information and advice. Several days after the *Origin* appeared, Darwin self-critically and half-jokingly acknowledged the enormous support that Gray had provided. He wrote to Gray on November 29, 1859: "This shall be an extraordinary note as you have never received from me, for it shall not contain one single question or request."[36]

Managing People

Darwin seemed particularly well suited by skill, by habit of mind, and by personality for the large management and organizational effort required to carry his species work forward. Good organizers and managers understand the meaning of time, how precious it is, and how to organize to take full advantage of it. Darwin had a healthy respect for time and an ability to organize the most effective use of it. "He never forgot how precious . . . [time] was . . .," Francis Darwin later remembered; "he would often say, that saving the minutes was the way to get work done . . . he never wasted a few spare minutes from thinking that it was not worth while to set to work."[37]

We have already noted Darwin's schedule at Down. It was an attempt to make full use of every minute in an attenuated work day. It represented a good manager's compromise between a desire for work and a need for rest. It also represented Darwin's overall desire to waste nothing, to leave no loose ends, to treat time as a high-priced commodity.

The same eager desire not to lose time is evident in his work. He moved quickly while experimenting, Francis Darwin tells us. He insisted that every experiment he performed was to be of some use. In the literary part of his work he had the same horror of losing time. Darwin was exasperated by the need to rewrite. His sensitivity to wasting time was such that he dreaded having to tell a story twice, or reread a book or journal article.[38]

Even his letters reflected his appreciation of and respect for time. Although full of the personal courtesies of the period, they were very businesslike and matter-of-fact. Darwin dispensed with the opening pleasantries as quickly and adroitly as possible in order to get to the real point of the letter: his need for a specimen, his desire for the answer to a question, an observation that he would like made.

Good organizers and managers also have personalities that make others want to work for them. Darwin had a charm, graciousness, and wit, really an irresistible, infectious personal vitality and magnetism, that drew people to him. His letters are replete with good humor, flattery, and friendly persuasion.[39] He seemed never able to thank others enough for their help, and he was always ready to offer his own help in return. "I should be a most ungrateful dog," he wrote to Asa Gray on August 11, 1858, "after all the invaluable assistance you have rendered me, if I did not do anything which you asked."[40] Sensitive to the feelings of others, he took great care in asking for assistance. "There is one other point on which I individually should be extremely much obliged," he typically wrote to Gray on May 2, 1856, "if you could spare the time to think a little bit and inform me. . . ."[41] Always the diplomat, Darwin went to great lengths to make people feel appre-

ciated for the help they provided. "How much he [Darwin] thought
of the trouble he gave others by asking questions, will be well
enough shown by his letters," Francis Darwin later remarked;
". . . it is difficult to say anything about the general tone of his
letters, they . . . speak for themselves. The unvarying courtesy of
them is very striking. I had proof of this quality in the feeling
with which Mr. Hacon, his solicitor, regarded him. He had never
seen my father, yet had a sincere feeling of friendship for him,
and spoke especially of his letters as being such as a man seldom
receives in the way of business:—'Everything I did was right, and
everything was profusely thanked for.' "[42]

Francis Darwin recalled that his father always made it clear in
correspondence with the poultry expert Tegetmeier that Tegetmeier's "knowledge and judgment were completely trusted and
highly valued by him. Numerous phrases, such as 'your note is a
mine of wealth to me,' occur, expressing his sense of the value of
Mr. Tegetmeier's unstinting zeal and kindness, or his 'pure and
disinterested love of science.' "[43]

There are many other examples of the profuse thanks Darwin
lavished upon those who were kind enough to assist him: To Fox,
on March 27, 1855, Darwin wrote, "I assure you I thank you heartily for your proffered assistance";[44] to Gray on June 8, 1855 he
wrote, "I thank you cordially for your remarkably kind letter of
the 22d ult., and for the extremely pleasant and obliging manner
in which you have taken my rather troublesome questions";[45]
and to Laurence Edmondston on May 3, 1856, "I beg to thank
you truly for your kind & very interesting answers to my queries."[46]

Part of the charm of Darwin's letters was the tone of meek hu-
mility, exaggerated self-denigration, and sincere apology that pervaded them—a tone, often spiced with wit, that made it difficult
for others not to like Darwin and not to want to help him if possible. To Hooker Darwin described himself as "an ungrateful dog"
for not answering one of his friend's letters sooner,[47] and as "the
most miserable, bemuddled, stupid dog in all England" for a
miscalculation Darwin made concerning the variation of large
genera.[48] He was Hooker's "insane and perverse friend,"[49] his "in-
tolerable but affectionate friend."[50] "Do be a good Christian and
not hate me," he kiddingly pleaded with Hooker after asking him
some questions.[51]

To Fox he was "your most troublesome friend."[52] "My correspondence has cost you a deal of trouble," he candidly admitted.[53]
And to Asa Gray he remonstrated: "How can I apologise enough
for all my presumption and the extreme length of this letter?"
while he added facetiously: "The great good nature of your letter
to me [which was full of information for Darwin's species work]
has been partly the cause, so that, as is too often the case in this
world, you are punished for your good deeds."[54]

Darwin's charming and gracious nature was also visible in his
works. "His courteous and conciliatory tone towards his reader
is remarkable," Francis Darwin has remarked; " . . . the reader
feels like a friend who is being talked to by a courteous gentleman, not like a pupil being lectured by a professor. The tone of
such a book as the 'Origin' is charming, and almost pathetic; it
is the tone of a man who, convinced of the truth of his own views,
hardly expects to convince others; it is just the reverse of the style
of a fanatic, who wants to force people to believe. The reader is
never scorned for any amount of doubt which he may be imagined to feel, and his skepticism is treated with patient respect."[55]

Darwin was as charming in person as he was in his letters and
his works. He was always a warm, vibrant, and captivating host.
" . . . The presence of visitors excited him, and made him appear
to his best advantage," Francis Darwin later remembered.[56] Darwin looked after his guests conscientiously; he was always animated and friendly in their presence. His manner was one of ease
and lack of pretense or affectation, and he spoke in a natural and
simple way. Perhaps most important to visitors, he was a good
listener.[57]

"He [Darwin] was particularly charming when 'chaffing' any
one, and in high spirits over it," Francis Darwin recalled; " . . .
his manner at such times was light-hearted and boyish, and his
refinement of nature came out most strongly. So, when he was
talking to a lady who pleased and amused him, the combination
of raillery and deference in his manner was delightful to see. . . .
When my father had several guests he managed them well, getting a talk with each, or bringing two or three together round his
chair."[58] In these conversations there was much fun and humor.
He always had fun with Huxley. But with Lyell and, as we have
seen, with Hooker, arguments could get heated, especially on topics of natural history. In any case, Darwin was the sort of person

for whom friends would go out of their way to provide assistance. "He undoubtedly had, to an unusual degree," Francis Darwin has written, "the power of attracting his friends to him."[59]

Good organizers and managers also have a determination and drive to get the job at hand done. They try to let nothing stand in their way. Darwin had an innate stubbornness and perseverance that drove him to get the information he needed. Darwin simply would not accept no for an answer when it came to requests regarding his species work.

Underlying his politeness and charm, his tone of courtesy and amiability, was a sternness and a calculated determination to get what he wanted. In a letter on May 17, 1855, mentioned above, Darwin reassured Fox: "You will hate the very sight of my handwriting, but after this time . . . I will ask for nothing more, at least for a long time."[60] Yet six days later Darwin asked Fox for, among other things, a one-week-old mongrel chicken.[61] To Gray he did not even attempt to hide his determination. On May 2, 1856, he wrote: "I am sure I have given you a larger dose of questions than you bargained for, and I have kept my word and treated you just as I do Hooker. Nevertheless, if anything occurs to me during the next two months, I will write freely, believing that you will forgive me and not think me very presumptuous."[62] His questions to Gray continued unabated.

The same determination and force of purpose were apparent in his experiments. Darwin had the ability, Francis Darwin remarked, to pursue a subject thoroughly; "he used almost to apologise for his patience, saying that he could not bear to be beaten, as if this were rather a sign of weakness on his part. He often quoted the saying, 'It's dogged as does it'; and I think doggedness expressed his frame of mind almost better than perseverance. Perseverance seems hardly to express his almost fierce desire to force the truth to reveal itself."[63]

Doubt and Uncertainty

There were certain aspects of Darwin's species work, however, that seemed unaffected by his considerable organizational and management abilities. One of these was an underlying doubt he seemed to feel about the value of his theory.

Early in Darwin's work on species, he seemed determined to present arguments both for and against the view of the common descent of species. The second part of his 1844 *Essay*, a preliminary, attenuated version of the *Origin*, was entitled "On the Evidence favourable and opposed to the view that Species are naturally formed races, descended from common Stocks."[64] There are also early indications in his correspondence of Darwin's desire to argue both sides of the issue. In an 1844 letter to Hooker, Darwin wrote: "I forget my last letter, but it must have been a very silly one, as it seems I gave my notion of the number of species being in great degree governed by the degree to which the area had been often isolated and divided; I must have been cracked to have written it, for I have no evidence, without a person be willing to admit all my views, and then it does follow; but in my most sanguine moments, all I expect, is that I shall be able to show even to sound Naturalists, that there are two sides to the question of the immutability of species; [one of them being] . . . that facts can be viewed and grouped under the notion of allied species having descended from common stocks."[65]

In a letter in 1845 to Leonard Jenyns, Darwin again indicated his intention of presenting both sides of the descent question: "In my wildest day-dream, I never expect more than to be able to show that there are two sides to the question of immutability of species, *i.e.*, whether species are *directly* created or by intermediate laws . . . that a collection of . . . facts would throw light either for or against the view of related species being co-descendants from a common stock."[66]

Ten years later Darwin's purpose had not changed. "I am hard at work at my notes collecting and comparing them," he wrote to Fox on March 19, 1855, "in order in some two or three years to write a book with all the facts and arguments, which I can collect, *for and versus* the immutability of species."[67] A week later, Darwin wrote again to Fox: "I forget whether I ever told you what the object of my present work is,—it is to view all facts that I can master . . . in Natural History (as on geographical distribution, palaeontology, classification, hybridism, domestic animals and plants, &c., &c., &c.) to see how far they favour or are opposed to the notion that wild species are mutable or immutable: I mean with my utmost power to give all arguments and facts on both sides."[68]

Darwin's desire to show both sides of the species question was indicative, I believe, of a pervasive uneasiness and persistent doubt about the value of his work on this subject—an uneasiness and doubt that continued to plague him in varying degrees up to the publication of the *Origin* in November of 1859.

We see this uneasiness as far back as his early letters on species written to Hooker. In these letters Darwin questioned the value of his work, or at least he seemed reluctant to admit it to his friend. In a letter to Hooker of January 11, 1844, cited above, Darwin revealed that since his return from the *Beagle* voyage he had been "engaged in a very presumptuous work, and I know no one individual who would not say a very foolish one. . . . I think I have found out (here's presumption!) the simple way by which species become exquisitely adapted to various ends."[69]

A few months later, Darwin wrote to Hooker to explain his view of the role of isolation in species development: "This will justly sound very hypothetical. I cannot give my reasons in detail; but the most general conclusion, which the geographical distribution of all organic beings, appears to me to indicate, is that isolation is the chief concomitant or cause of the appearance of *new* forms. . . . But such speculations are amusing only to one self, and in this case useless, as they do not show any direct line of observation: if I had seen how hypothetical [is] the little, which I have unclearly written, I would not have troubled you with the reading of it." Significantly, Darwin ended the letter: "Believe me, —at last not hypothetically, Yours very sincerely, C. Darwin."[70]

In another letter to Hooker in September 1845, also cited above, Darwin talked of his "long self-acknowledged presumption" of working on his species theories without having worked out his "due share of species." "But now for nine years," he added somewhat wistfully, "it has been anyhow the greatest amusement to me."[71] In 1849, Darwin wrote to Hooker about his "rude species-sketch." If it had any small role in helping Hooker solve a problem, "it has already done good and ample service, and may lay its bones in the earth in peace."[72]

In 1854, following completion of his work on Cirripedes—work that had occupied the greatest part of his attention during the late 1840s and early 1850s—Darwin prepared to return full-time to his species theory, though, as he indicated in a letter to Hooker on March 26, 1854, he did so with considerable trepidation: "How awfully flat I shall feel, if when I get my notes together on species, &c., &., the whole thing explodes like an empty puff-ball."[73]

Throughout 1855, doubt and uncertainty continued to plague him. In a letter to Hooker in July, Darwin revealed that he would not feel so uncomfortable about the burden he was placing on his friend if he could feel more certain about the value of his ideas: "I should have no scruple in troubling you if I had any confidence what my work would turn out. Sometimes I think it will be good; at other times I really feel as much ashamed of myself as the author of the *Vestiges* ought to be of himself."[74] And on October 12, 1855, Darwin wrote to Leonard Jenyns that he felt he was "a bold man to lay . . . [himself] open to being thought a complete fool, and a most deliberate one" by his continued work on species.[75]

By 1856, Darwin seemed to have gained a little more confidence in his ideas. The problem still remained whether he could convince any others of their value. On July 20, 1856, he first described his species theory to Gray and argued that his ideas seemed "to explain too much, otherwise inexplicable, to be false," even though he did not expect Gray to accept them. "I know that this [these ideas] will make you despise me."[76] Ten days later, in a letter to Hooker, he expressed some hope that many of the more serious problems associated with his species theory had been removed, though, doubtful again, he added: "God knows it may be all hallucination."[77]

By August, Darwin seemed optimistic again, at least in his own view of his ideas, though he was still concerned about what others might think. He wrote to Hooker about his variable feelings: "Sometimes I am in very good spirits and sometimes very low about it [his species work]. My own mind is decided on the question of the origin of species; but, good heavens, how little that is worth! . . . "[78]

Throughout 1858 and 1859, until the publication of the *Origin*, Darwin's conviction of the worth of his species work continued to waver. "I must come to some definite conclusion whether or not entirely to give up the ghost [about my species theory]," he somberly admitted to Hooker in a letter dated February 28, 1858.[79] In a letter to Asa Gray of August 11, 1858, Darwin attempted to use Hooker's recent "conversion" to his ideas as an argument against Gray's doubts: "I cannot give you facts," Darwin wrote, "and I must write dogmatically, though I do not feel so on any point. I may just mention, in order that you may believe that I have *some* foundation for my views, that Hooker has read my MS., and though he at first demurred to my main point, he has since told me that further reflection and new facts have made him a convert."[80]

By early 1859, Darwin had adopted this reasoning regarding publication of his species theory: the *Origin* might not be successful; there is a good chance that its arguments will not convince a wide range of people; yet, if the work helps to carry the theory of descent with modification by means of natural selection a few steps forward, then it can be considered to have made a reasonable contribution. He wrote to Hooker on January 20, 1859: "I always comfort myself with thinking of the future, and in the full belief that the problems which we are just entering on, will some day be solved; and if we just break the ground we shall have done some service, even if we reap no harvest."[81]

A few months later Darwin was still concerned about his species ideas, particuarly about his ideas on geographical distribution. As previously mentioned, he wrote to Hooker on March 2, 1859, concerning his draft *Origin* chapter on the subject: "I should like you much to read it. . . . I want it, because I here feel especially unsafe, and errors may have crept in."[82] In a letter of March 15, 1859, Darwin added: "P.S. You cannot tell what a relief it has been to me your looking over this chapter, as I felt very shaky on it."[83]

On the eve of the publication of the *Origin*, Darwin still seemed to be wrestling with the question of the quality of his work. To whom should he send copies? He seemed to want to send the *Origin* only to those foreign naturalists of a "speculative" bent. Groping, he wrote to Hooker on October 15, 1859: "Do you know any philosophical botanists on the Continent, who read English and care for such subjects?"[84] In a letter written on the same day to Huxley, Darwin asked: "Can you tell me of any good and *speculative* foreigners to whom it would be worthwhile to send copies of my book, on the 'Origin of Species'?"[85]

A Variety of Problems

Why was Darwin so uncertain? One problem, discussed briefly in the last chapter, was his inability to explain his ideas adequately. We have seen Hooker's problem with fully understanding them. Huxley did not seem to grasp Darwin's ideas much better, even after the *Origin* had appeared and Huxley had become one of Darwin's chief advocates. As Darwin wrote to Hooker on February 14, 1860, concerning Huxley's lecture explicating the theory of natural selection: "After conversation with others and more reflection, I must confess that as an exposition of the doctrine . . . [Huxley's] lecture seems to me an entire failure. I thank God I did not think so when I saw Huxley; for he spoke so kindly and magnificently of me, that I could hardly have endured to say what I now think. He gave no just idea of Natural Selection."[86]

Darwin himself seemed to have trouble fully grasping his ideas. His quick assembly of them in a shortened form in 1858 and 1859 was a blessing in disguise. "You cannot imagine what a service you have done me in making me make this Abstract," he wrote to Hooker on October 6, 1858, "for though I thought I had got all clear, it has clarified my brains very much, by making me weigh the relative importance of the several elements."[87]

Another problem was that the deeper Darwin got into his subject, the more lost he felt. The work he had undertaken seemed, at times, to overwhelm him: To Fox he wrote: "I often doubt whether the subject will not quite overpower me" (March 27, 1855);[88] to Wallace: "I am now preparing my work for publication, but I find the subject so very large. . . . " (May 1, 1857);[89] to Sym Covington: "This work will be my biggest . . . I have to discuss every branch of natural history, and the work is beyond my strength and tries me sorely" (May 18, 1858);[90] and to Asa Gray: "I find, alas! each chapter takes me on an average three months, so slow I am. There is no end to the necessary digressions. I have just finished a chapter on Instinct, and here I found grappling with such a subject as bees' cells, and comparing all my notes made during twenty years, took up a despairing length of time" (April 4, 1859).[91]

At one point, in what to Darwin had become the treacherous process of preparing the *Origin* for publication, he sought refuge from the pressures of his work in a period of hydropathy in the country. His health had been deteriorating. From Moor Park in April 1858, he wrote an idyllic letter to his wife, Emma, in which he poetically summed up his feelings at the time: "Yesterday . . . I strolled a little beyond the glade for an hour and a half, and enjoyed myself—the fresh yet dark-green of the grand Scotch firs, the brown of the catkins of the old birches, with their white stems, and a fringe of distant green from the larches made an excessively pretty view. At last I fell fast asleep on the grass, and awoke with a chorus of birds singing around me, and squirrels running up the trees, and some woodpeckers laughing, and it was as pleasant and rural a scene as ever I saw, and I did not care one penny how any of the beasts or birds had been formed."[92]

Another problem was Darwin's fear that people would think his work unprofessional because, in a shortened version of his originally intended work, he would be unable to provide adequate scholarly references. This fear arose after 1856 from two separate, though related, events. The first was the prophetic suggestion by Lyell in the spring of 1856 that Darwin prepare and publish a short sketch of his views to avoid the possibility of losing priority. To this suggestion, Darwin reacted with apprehension. He wrote to Lyell on May 3, 1856: " . . . with respect to your suggestion, I hardly know what to think . . . but it goes against my prejudices. To give a fair sketch would be absolutely impossible, for every proposition requires such an array of facts . . . if I did publish a short sketch, where on earth should I publish it?"

It will be simply impossible for me to give exact references; anything important I should state on the authority of the author generally; and instead of giving all the facts on which I ground my opinion, I could give by memory only one or two. In the Preface I would state that the work could not be considered strictly scientific, but a mere sketch or outline of a future work in which full references, &c., should be given. Eheu, eheu, I believe I should sneer at any one else doing this. . . . [93]

To Hooker a few days later (May 9, 1856) Darwin complained:

I positively will *not* expose myself to an Editor or a Council, allowing a publication for which they might be abused. If I publish anything it must be a *very thin* and little volume, giving a sketch of my views and difficulties; but it is really dreadfully unphilosophical to give a *resumé* without exact references, of an unpublished work."[94]

The second event was the letter from Wallace in June 1858 containing Wallace's theory of descent by natural means of selection, which almost exactly duplicated Darwin's own theory on the subject. This letter led Darwin to abandon the larger work with which he was then occupied in favor of a shorter abstract which eventually became the *Origin of Species*. Darwin wrote despairingly to Lyell on July 18, 1858 about his new, forced task: "I am going to prepare an . . . abstract; but it is really impossible to do justice to the subject, except by giving the facts on which each conclusion is grounded, and that will, of course, be absolutely impossible."[95]

Yet another problem was Darwin's extreme personal sensitivity to negative public reaction that might greet his theories and, he feared, diminish his status within his profession. From the beginning of his work on species, he was particularly concerned with what other people might think of his ideas. He wrote pessimistically to Hooker in October of 1846: "I am going to begin some papers on the lower marine animals, which will last me some months, perhaps a year, and then I shall begin looking over my ten-year-long accumulation of notes on species and varieties, which, with writing, I dare say will take me five years, and then, when published, I dare say I shall stand infinitely low in the opinion of all sound Naturalists—so this is my prospect for the future."[96]

After the publication of the *Vestiges of Creation* in 1844,[97] the negative, and at times abusive, reaction to that work served as a frightening example to Darwin of what might befall his advocacy of a theory of descent with modification. Darwin revealed to Hooker in September 1849 that "though I shall get more kicks than half-pennies, I will, life serving, attempt my work" but added that he hoped someday future naturalists would not scorn him as they had Lamarck and the anonymous author of the *Vestiges*: "Lamarck is the only exception . . . of an accurate describer of species . . . in the Invertebrate Kingdom, who has disbelieved in permanent species, but he in his absurd though clever work has done the subject harm, as has Mr. Vestiges, and, as (some future loose naturalist attempting the same speculations will perhaps say) has Mr. D. . . . ."[98] To Huxley, who had brutally criticized the *Vestiges*, a displeased Darwin wrote on September 2, 1854, as much in self-defense as in fairness: "I cannot think but that you are rather hard on the poor author. I must think that such a book, if it does no other good, spreads the taste for Natural Science. . . . But I am perhaps no fair judge, for I am almost as unorthodox about species as the *Vestiges* itself, though I hope not . . . so unphilosophical."[99]

Darwin's sensitivity to possible negative public reaction naturally increased as the time approached for publication of the *Ori-*

gin. Throughout 1858 and 1859, his letters reflect a burgeoning anxiety. As we saw in the last chapter, Darwin implored Hooker in a letter of October 12, 1858 "not to pronounce too strongly against Natural Selection, till you have read my abstract, for though I dare say you will strike out *many* difficulties, which have never occurred to me; yet you cannot have thought so fully on the subject as I have."[100] In a follow-up letter to Hooker the next day, Darwin apologized for the request, explaining: "But the truth is, that I have so accustomed myself, partly from being quizzed by my non-naturalist relations, to expect opposition and even contempt, that I forgot for the moment that you are the one living soul from whom I have constantly received sympathy."[101]

Several months later, on December 24, 1858, Darwin expressed his fear to Hooker that religious objections might be raised against his ideas: "The subject [natural selection] really seems to me too large for discussion at any Society, and I believe religion would be brought in by men whom I know."[102] He was also fearful that the public might not find the *Origin* very interesting or understandable. He predicted to Hooker on June 22, 1859: "The public will find it [i.e., the *Origin*] intolerably dry and perplexing."[103]

He was already, in a letter to Wallace of August 9, 1859, predicting Richard Owen's attack: "Owen, I do not doubt, will bitterly oppose us. . . . "[104] In a letter to Hooker of October 13, 1858, Darwin recalled that he had been warned several years before by his old friend Falconer that his ideas would "do more mischief than any ten other naturalists would do good" and that he had already half-spoiled Hooker.[105] Soon after the publication of the *Origin*, in a letter of December 17, 1859, Darwin half-humorously confronted Falconer: "You . . . [are] very antagonistic to my views on species. I well knew this would be the case. I must freely confess, the difficulties and objections are terrific; but I cannot believe that a false theory would explain, as it seems to me it does explain, so many classes of facts. Do you ever see Wollaston? He and you would agree nicely about my book—ill luck to both of you."[106]

Possible Converts

Darwin had, in fact, many realistic fears to contend with. At the end, he had very little backing for his ideas. It is clear that he expected opposition at least from Owen, Falconer, and Wollaston.

Aside from Wallace, who had independently arrived at the same ideas, only Hooker could be firmly counted as a convert, and Hooker had converted only quite recently—perhaps as early as 1858 (at which time Darwin wrote to Gray about Hooker's "conversion")[107]—more likely in the spring of 1859 (Darwin wrote to Wallace on April 6th: "I forget whether I told you that Hooker, who is our best British botanist and perhaps the best in the world, is a full convert . . . ")[108]—certainly by the fall ("Hooker has come round," Darwin wrote to Fox on September 23, 1859, "and will publish his belief soon.").[109]

The great majority of Darwin's naturalist friends and acquaintances remained undecided. Darwin hoped that Gray would "go round," he wrote to Hooker on May 11, 1859, "for it is futile to give up very many species, and stop at an arbitrary line at others."[110] But there was no indication one way or the other from Gray before the publication of the *Origin.* I do not think that Darwin made any serious attempt to convert Fox. "I shall be curious to hear what you think of it, but I am not so silly as to expect to convert you," he wrote to Fox on September 23, 1859.[111]

Lyell caused Darwin much anxiety. In the letter to Fox just quoted, Darwin seemed optimistic about Lyell coming over: "Lyell has read about half of the volume in clean sheets, and gives me very great *kudos.* He is wavering so much about the immutability of species, that I expect he will come round."[112] About a month later, in a letter to Hooker on October 15, 1859, Darwin was not so certain. He was concerned that Lyell was upset about the lengths to which Darwin had gone in the *Origin:* "Lyell is going to reread my book, and I yet entertain hopes that he will be converted, or perverted, as he calls it," Darwin wrote.[113] Ten days later Darwin was optimistic about Lyell again. Hooker had written him that Lyell was closer to Darwin's position on species than Darwin had thought. "What you say about Lyell pleases me exceedingly . . .," he wrote to Hooker; "I had not at all inferred from his letters that he had come so much round. . . . "[114] In the end, Lyell, disturbed at the implications Darwin's theories had for the origin of man, remained unconverted.[115]

Huxley seemed to Darwin a very unlikely convert at the time. At one point early in 1859, Darwin was hopeful, though still uncertain, of Huxley's allegiance. He wrote to Wallace on April 6, 1859: "Huxley is changed, and believes in mutation of species:

whether a convert to us, I do not quite know."[116] A few months later, in a letter to Huxley on October 15, 1859, Darwin was less hopeful: "I shall be *intensely* curious to hear what effect the book produces on you. I know that there will be much in it which you will object to, and I do not doubt many errors. I am very far from expecting to convert you to many of my heresies. . . . "[117] A week later, on October 23, 1859, Darwin wrote to Hooker: "I am intensely curious to hear Huxley's opinion of my book. I fear my long discussion on Classification will disgust him; for it is much opposed to what he once said to me."[118] On the eve of the publication of the *Origin* Darwin still was uncertain: "I long to learn what Huxley thinks," he wrote to Hooker in November 1859.[119]

Alone on the Eve

In a situation in which one senses that widespread support is not forthcoming, there is a tendency, I believe, to value the support of a few as if it were equal to the support of many. Darwin seemed to be in such a situation. Feeling that the general public and most of his fellow naturalists might ridicule him, he took the position, which he seems to have rationalized with at least some success, that the support of a few key people was better than the support of the majority. To Huxley in October 1859, Darwin wrote: "If, on the whole, you and two or three others think I am on the right road, I shall not care what the mob of the naturalists think."[120] And to Hooker in November of 1859: "If some four or five *good* men come round nearly to our view, I shall not fear ultimate success."[121]

Yet, it is doubtful if he had even four or five supporters. The *Origin* was ultimately a leap in the dark for Darwin—a courageous act and, in some people's eyes, especially Darwin's, a premature one. But after Wallace, there was no time to worry. The *Origin* appeared on November 24, 1859.

6

The *Origin* as the Least Objectionable Theory

Preparing the Argument

Weaknesses in the Argument

Another problem confronted Darwin that must have added to his anxiety concerning publication of his ideas. This problem went beyond his failure to communicate his theory clearly, or his concern over the overwhelming scope of his work, or his inability to provide scholarly references, or his lack of adherents; it concerned the argument of the *Origin* itself. Though often brilliantly and ingeniously composed, his argument was based, in many instances, on new and often unsubstantiated hypotheses, sometimes fuzzy analogies and metaphors, the repudiation of competing explanations, and a frequent plea to complexity and general ignorance, rather than on compelling, clearly incontrovertible evidence in its own support; and it is clear that Darwin knew this.[1]

In the end, in addition to attempting to demonstrate that his theory was right, which Darwin certainly tries to do, he is forced, in an important part of his argument in the *Origin*, to assert that other theories attempting to explain the same phenomena are wrong. Darwin is thus confronted with the problem of propounding not necessarily the correct theory, but the least objectionable one—a theory that, in terms of tangible evidence, was really not much stronger than those he was attempting to refute.

Theoretical Concerns: The Transmutation Notebooks

Darwin was aware of various potential weak spots in his theory, even before he had read Malthus. He knew that many important aspects of his argument lacked key supporting evidence.

101

Throughout the Transmutation Notebooks we see at times a nervous and rather anxious Darwin attempting to cope with a variety of substantive problems.

Darwin felt, for example, that to explain the extinction of the South American quadruped was "difficult on any theory," including his own.[2] He expected "vast opposition" to his theory "on all subjects of classification."[3] The question of the evolution of instincts presented numerous problems for him. "Is not squirrel hoarding & killing grain acquirable through hoarding from short time?" he asked at one point. He was not sure, yet he was determined that his theory "must encounter all . . . difficulties" related to the question of instincts.[4]

The tricky subject of spontaneous generation he seems simply to have omitted. "My theory leaves quite untouched the question," Darwin wrote uncomfortably,[5] while what seemed to be the selected immutability of some species left him baffled. "Those species which have long remained are those . . . which have wide range and therefore cross and keep similar, . . . " he wrote; "this is difficulty: this immutability of some species."[6]

Other phenomena of nature presented problems for him. He felt that the recognized disparity between the mind of man and animals was an issue of considerable magnitude, even though he had shown a unity of type in the physical structure of both;[7] he knew that his theory could not explain the frequent lack of change in fossils in thick sedimentary beds. "My very theory requires each form to have lasted for its time," he wrote, "but we ought in same bed if very thick to find some change in upper & lower layers . . . good objection to my theory: a modern bed at present might be very thick & yet have same fossils. . . . "[8]

One of the most significant problems with which he had to deal concerned the lack of intermediate organic forms. The basic problem was that if organisms had evolved over millions of years in a slow, steady progression of forms, why do we not see innumerable intermediate forms everywhere, or at least the fossil remains of such forms? Darwin stated the problem at one point: ". . . it may be argued against theory of changes that if so, in approaching desert country or ascending mountain you ought to have a gradation of species, now this notoriously is/not the case."[9] Or, as he described the problem at another point: "—Why to be sure [if] there were a thousand intermediate/forms . . . [of the otter].—opponent will say: show them me."[10]

Several interrelated problems bore closely on the intermediate forms question. One was that if indeed intermediate forms did exist, what did these forms look like and how did they survive before they reached their present recognizable form? What, for example, did an otter look like before it became an otter? Or, a question of more difficulty, what was the eye before it became an eye? Darwin's theory applied not only to organisms but to organs as well. Darwin reluctantly concluded that we may never be able to trace the steps by which the organization of the eye passed from simpler to more complex forms. ". . . This really perhaps greatest difficulty to whole theory," he wrote with obvious concern.[11]

Another problem was that if intermediate forms did exist, by what process did one form change into another? In other words, *how* did specific forms actually evolve in nature? "People will argue and fortify their minds with such sentences as 'oh turn a Buccinum into a Tiger,' " Darwin wrote, "but perhaps I feel the impossibility of this more than any one. . . . "[12] Darwin reasoned that it would "be easy to prove persistent varieties in wild animals," but wondered "how to show species." He added rather anxiously: "I fear argument must rest upon analogy [from domestic breeding]. . . . "[13] That is, he feared he could really provide only indirect evidence for this idea.

Still another problem concerned the time required to effect various changes in organic forms. If species changed by gradual, almost imperceptible steps, one would have to assume a process of change involving the multiplication of small means over long periods of time. This was conceivable if one accepted the greatly extended period of time for earth history described by Lyell in his *Principles of Geology,* but Darwin realized that not everyone did. He sensed that without such an acceptance, the idea of small changes adding up to great effects would be rather difficult to explain. It contradicted common sense. "This multiplication of little means & bringing the mind to grapple with great effect produced," Darwin wrote, "is a most laborious & painful effort of the mind . . . & will never be conquered by anyone . . . who just takes up & lays down the subject without long meditation. . . . "[14] Darwin realized "the difficulty of multiplying effects" and knew that "to conceive the results with that clearness of conviction absolutely necessary as the basal foundation stone of further inductive reasoning" was very difficult without a recognition of the facts of modern geology (that is, Lyellian geology). "It is curious that geology by giving proper ideas of these subjects should be *absolutely* necessary to arrive at right conclusion about species," he concluded.[15]

Strategic Concerns: The Transmutation Notebooks

At the same time (1837–39) that Darwin was struggling with a series of substantive problems surrounding his theory, he was equally concerned with a variety of strategic questions directly related to the communication and acceptance of his ideas. Darwin sensed very clearly the difference between believing his theory was true and being able to prove that it was.[16]

Darwin's concern frequently took the form of instructions to himself on how best to argue his case, as if he were preparing himself for some great debate, or barrage of criticism, to come.[17] Regarding intermediate forms, for example, Darwin instructed himself to "quote *in detail* some good instance" existing in nature.[18] Concerning the process of slow change over millions of years, Darwin directed himself to "argue the case theoretically if animals did change excessively slowly whether geologists would not find fossils as they are."[19] Regarding variation, Darwin was concerned that although he had borrowed the mechanism of evolution from Malthus, he did not agree with the limits Malthus put on the extent to which variations could go. He wrote: "It may be said that wild animals will vary according to my Malthusian views, within certain limits, but beyond them not. . . . " "Argue against this . . . ," he instructed himself.[20] Many other examples could be cited.[21]

In all this, Darwin was concerned with arguing his case as openly as possible, that is, with delineating the problems and difficulties for everyone to see. If he did not know the answer to a difficult problem, he believed that the best strategy was simply to admit that he did not know. Regarding how the selection process works in nature, for example, Darwin advised himself to "make the difficulty apparent by cross-questioning" and then "give my theory."[22] Concerning the various means by which organisms can be transported, Darwin felt "if some [means of transport] cannot be explained more philosophical to state we do not know how transported."[23]

Darwin was also concerned about falling into unphilosophical pitfalls and about the impact that these would have on the strength of his argument. He felt, for example, that he should avoid going back to the first origin of things in his speculations, for in order to do so it would "be necessary to show how the first eye is formed," something which he knew he could not do.[24]

He was wary, in general, of too much speculation, too much hypothesizing. "Is there some law in nature [whereby] an animal may acquire organs, but lose them with more difficulty . . . such law would explain every thing," Darwin wrote. "*Pure hypothesis be careful . . . ,*" he immediately warned himself.[25] He found the "method of generalizing without tables of references highly unphilosophical."[26] Darwin's preferred method was to "work out hypothesis & compare it with results." If he acted otherwise, he felt his "premises would be disputed."[27]

Darwin's concentration on the strategic aspects of his argument was related to his concern with the value of his ideas—the nature and extent of his contribution. He was concerned with just how original his theory was and how it compared to similar theories attempting to explain the same phenomena. This concern seemed to be part of his overall feeling of doubt and uncertainty about the value of his theory. He felt he should read Aristotle to see if any of his views had precedent.[28] Recognizing what von Buch, Humboldt, Saint-Hilaire, and Lamarck had already written, Darwin admitted that he could "pretend to no originality of idea" and felt that the best strategy would be to place his lack of originality boldly before the reader, as he would his discussion of problems and issues. "State broadly scarcely any novelty in my theory," he instructed himself.[29]

While pretending to no great originality, he did seem determined to put some distance between his theory and those of his predecessors. With regard to his grandfather Erasmus Darwin's theory, he advised himself to "say my grandfathers [sic] expression of generat[ion] being highest end of organization *good expression* but does not include so many facts as mine."[30] There are numerous references in the notebooks in which Darwin attempts to distinguish clearly his theory from Lamarck's: " . . . changes not result of will of animals, but law of adaptation as much as acid and alkali;[31] "my theory very distinct from Lamarck's";[32] "with respect to how species are [formed], Lamarck's 'willing' doctrine absurd";[33] "Lamarck's willing . . . not applicable to plants."[34]

Most important, Darwin seemed concerned with elucidating the differences between his theory and the Creationist view, in a way, of course, favorable to his theory. The Creationist theory was Darwin's chief competitor; it was the theory of the natural world most generally accepted by British scientists of the period, and hence the theory that Darwin would need to discredit if his own were to gain acceptance.

Darwin's main argument against the Creationist position was that it simply did not explain much or make logical sense. Darwin found the useless wings under the elytra of beetles, for example, inconsistent with the theory of creation—"if simple creation, surely would have [been] born without them."[35] Darwin also had trouble understanding how a Creator could have devised mental faculties for so many diverse organisms. If someone were to ask him by what power the Creator was able to do that, Darwin facetiously admitted: "I will confess my profound ignorance."[36] Furthermore, Darwin found it difficult to understand how or why the Creator should be involved in determining the most peculiar instincts of the lowliest organisms. "It surely is not worth interposition of deity to teach squirrel to kill ears of corn," Darwin argued.[37]

Darwin found the Creationist position perhaps least illuminating, least "logical," when it tried to explain geographical distribution. He wondered whether it would be logical to assume that the Creator would actually make plants to adorn a newly arisen volcanic island in the ocean. Was all this very probable?[38] Was it very likely, he asked, that the Creator made a rat especially for the island of Ascension? Darwin advises himself here to "argue the case of probability," meaning, presumably, raise the issue of whether this is likely to have occurred.[39] Darwin also wondered why "wandering birds such [as] sandpipers not new at Galapagos. Did the creative force know these species could arrive. — did it only create those kinds not so likely to wander — did it create two species closely allied to Mus[cicapa] coronata/but not coronata. . . ."[40]

Darwin belittled the idea, which he attributed to the Creationists, that certain facts of nature which did not fit neatly into a Creationist viewpoint were there to "fool" man—"as likely as fossils in old rocks for same purpose!," Darwin wrote.[41] To Darwin the existence of "fresh creations is mere assumption, it explains nothing further."[42]

Darwin also saw the advantage of simplicity his theory had over the Creationist view. "Astronomers might formerly have said that God ordered each planet to move in its particular destiny," Darwin wrote, "but how much more simple and sublime power let attraction act according to certain law, such are inevitable consequences. . . ."[43] Surprisingly enough, his theory even seemed to give greater grandeur to God. Here he was attacking the Creationist position at its very heart. "Has the Creator since the Cambrian formation gone on creating animals with same general structure.— Miserable limited view . . . ," Darwin argued, " . . . how far grander [is Darwin's theory] than idea from cramped/ imagination that God created . . . the Rhinocerous of Java & Sumatra, that since the time of the Silurian he has made a long succession of vile molluscous animals. How beneath the dignity of him. . . ."[44]

Finally, for Darwin, there were simply too many facts of nature—usually dealing with the geographical distribution of organisms—that only his theory could explain. As large mammalia are not found on all islands, Darwin wrote: " . . . if act of fresh creation, why not produced on New Zealand; if generated, an answer can be given. . . ."[45] Darwin questioned the relationship of organic forms on neighboring continents: "Did Creator make all new, yet forms like neighbouring continent . . . my theory explains this but no other will."[46] As we shall see, the theme of Darwin's theory being able to explain various facts of nature which the Creationist view could not is developed quite powerfully in the *Origin*.

Darwin felt that the only proper basis for understanding the natural world was a view of nature deeply grounded in the lawlike nature of the universe. He felt that this emphasis on law distinguishd his theory from the Creationist view, and he believed it was an important criterion by which one should judge the usefulness of his theory. "The grand question which every naturalist ought to have before him when dissecting a whale, or classifying a mite, a grampus or an insect is What are the Laws of Life?," he proposed.[47] Yet he was concerned about the negative and materialistic implications of too great an emphasis on law. "Why is thought being a secretion of brain, more wonderful than gravity a property of matter?," Darwin asked. But then he quickly chided himself: "Love of deity effect of organization, oh you materialist!"[48]

Perhaps as a compromise, Darwin maintained a sort of deistic view of the development and workings of the natural world, a view which was careful not to emphasize law to the exclusion of all else. Darwin contended that although the universe might have been created by God and God might have created the laws by which it ran, He no longer played an active role in its operations; secondary laws determined the nature of the universe after the first creation. Darwin came to see this emphasis on secondary causes as a possible key in understanding the origin of species.[49]

The M and N Notebooks

In the so-called M and N notebooks,[50] begun while he was still filling the Transmutation Notebooks with thoughts (Darwin began the M Notebook at the same time—July 1838—as the third Transmutation Notebook), Darwin's interests shift a bit, and he seems more concerned with the implications of his theory—how far it can be applied to instincts and mental activities, its impact on other fields of study—and less so with substantive problems or with strategic problems in getting his theory communicated and accepted. Given Darwin's general interests at the time, however, there is evidence of some carryover of concern in the M and N notebooks with substantive and strategic problems.

Darwin is still concerned with difficulties surrounding his theory and how he might best handle them. "Nearly all will exclaim, your arguments are good but look at the immense difference between man, . . . judge only by what you see . . . ," Darwin wrote considering one common-sense argument that might be used against his theory. Then he added, as a possible rejoinder: "Compare the Fuegian & Ourang-Outang, & dare to say differences so great. . . . "[51]

Darwin is also concerned with the strategy of arguing his case and continues to instruct himself on the matter. He feels with heredity, for example, "the real argument fixes on hereditary disposition & instincts" and orders himself to " . . . put it so."[52] Regarding the memory of good times and pleasures, Darwin advises himself to " . . . begin discussion — by saying what is Happiness? — When we look back to happy days, are they not those of which all our *recollections* are pleasant."[53] Or, concerning using the development of language as a supporting analogy for his arguments in favor of progressive development, Darwin reasons: "I may put the argument that many learned men seem to consider there is good evidence in the structure of language, that it was progressively formed. . . ."[54]

Darwin is also once again concerned with the question of his originality, with the uniqueness of his theory, and with distinguishing his ideas from those of others. Lamarck, for example, seems, from Darwin's point of view, to start where Darwin had started; but Lamarck ends up with "no facts," with a theory "mingled with much hypothesis."[55]

Darwin again attacks the Creationist position, this time relating his criticism to his vision of a lawlike nature, with continuing emphasis on God as the original Creator and the importance of secondary laws. "Those savages . . . make the same mistake . . . as does that philosopher who says the innate knowledge of creator [is] . . . implanted in us . . . by a separate act of God," Darwin wrote, "& not as a necessary integrant part of his most magnificent laws, which we profane in thinking not capable to produce every effect of every kind which surrounds us."[56] Darwin felt that this "unwillingness to consider [the] Creator as governing by laws is probable that as long as we consider each object an act of separate creation, we admire it more, because we can compare it to the standard of our own minds, which ceases to be the case when we consider the formation of laws invoking laws & giving rise at last even to the perception of a final case."[57] He was amazed that while man "can allow/satellites/plants, sun, universes, nay whole systems of universes to be governed by laws, [yet] . . . the smallest insect, we wish to be created at once by special act, provided with its instincts, its place in nature, its range. . . ."[58]

Finally, in the M and N notebooks we see a new strong emphasis on the materialist nature of the universe and a concomitant heightened concern with the atheistic implications of such views. Concerning materialism, Darwin felt that free will was related to mind the same way chance was to matter,[59] that " . . . free will & chance are synonymous—Shake ten thousand grains of sand together & one will be uppermost — so in thoughts, one will rise according to law."[60] "My wish to improve my temper," Darwin reasoned, "what does it arise from, but organization, that organization may have been affected by circumstances & education & by the choice which at that time organization gave me to will —

Verily the faults of the fathers, corporeal & bodily, are visited upon the children. . . . The above views would make a man a predestinarian of a new kind, because he would tend to be an atheist."[61] Darwin, the perceptive strategist, warned himself about expressing such views too openly: "To avoid stating how far, I believe, in Materialism, say only that emotions, instincts degrees of talent, which are hereditary are so because brain of child resembles parent stock."[62]

The 1842 and 1844 Essays

The last entry in Darwin's N Notebook is dated April 20, 1840. In June 1842, as we saw, he first attempted to put his theory on paper. It took the form of a 35-page sketch. In the summer of 1844, he expanded the effort to 230 pages.

These 1842 and 1844 sketches or *Essays*,[63] as I will call them, in some ways go much beyond the Transmutation and M and N notebooks and are like the *Origin* both in their argumentative form and in their use of similar language and sentence structure.

The *Essays* and the *Origin* use a similar form of argument; both begin with a description of the mechanism of evolution, a discussion of variation, and a description of selection. Both move from an analysis of domestic forms to a consideration of animals and plants in a state of nature. Both then discuss the difficulties facing Darwin's theory; then each deals with the special problem of instincts.[64]

There are several passages in the *Origin* which first appeared in almost the exact form in the *Essays* over fifteen years earlier. For example, the last sentence in the *Origin*—"There is grandeur in this view of life . . . "—is similar to the last sentence in both the 1842 and 1844 *Essays*;[65] the simile "We no longer look at an organic being as a savage does at a ship" is in the 1842 *Essay*, the 1844 *Essay*, and the *Origin*;[66] the sentence "Geology loses glory from imperfection of its archives, but it gains in the immensity of its subject" can be found in 1844 *Essay* and in a slightly different form in the *Origin*.[67] Other examples could be cited.[68]

In other ways, however, the *Essays* are not much different from the notebooks. As a finished product, for example, the *Essays* do not go much beyond the rough draft state of both the Transmuta-

tion and M and N notebooks. The 1842 *Essay*, Francis Darwin
tells us, was written on bad paper with soft pencil. In many places
it was extremely difficult to read. It was apparently written rapidly, with many erasures and corrections, and it did not seem to
have been reread with any care. "The whole is more like hasty
memoranda of what was clear to himself, than material for the
convincing of others."[69]

The 1844 *Essay* was not in much better shape. Darwin thought
of publishing it only as an "undesirable" expedient, and he wrote
it from memory without consulting any works. Although the 1844
Essay is more finished than the 1842 *Essay*, Francis Darwin felt
it too resembled an uncorrected manuscript.[70]

The notebooks and *Essays* are also very similar in their evidence
of Darwin's continuing concern with substantive and strategic
problems surrounding his theory. Regarding perfect organs, for
example, Darwin felt that the formation of such organs as the
eye and the ear could perhaps never be explained and that this
"first appears monstrous and to (the) end appears difficulty."[71]
On the possibility of multiple centers of creation, Darwin wrote
in the margin of one page of his 1844 *Essay:* "If same species appear
at two spot at once, fatal to my theory. . . . "[72] Regarding geographical distribution, he noted tersely, again in a margin of his 1844
Essay: "All this requires much verification."[73] Darwin admitted
concern over W. S. Macleay's Quinerian theory: "I discuss this
because if Quinarism true, I false."[74] And regarding the question
of the mutability of species, Darwin reasoned: "Besides other difficulties . . . difficulty when asked *how* did white and negro become altered from common intermediate stock: no facts."[75]

Darwin had two special problems with variations and variability: first, the problem of whether or not variations actually existed
in nature; second, the limits to variability—how far organisms
could vary. As we have seen, Malthus set definite limits to variation, limits which Darwin did not wish to accept.[76]

Darwin seemed somewhat uncertain whether variations actually existed in nature. At one point he admits: "Most organic
beings in a state of nature vary exceedingly little."[77] At other
points he is willing to argue for variation at least as a probability
in nature: "I have endeavoured to show that . . . variation or speci-
fication is not impossible, nay, in many points of view is absolutely probable . . . ";[78] " . . . it is highly probable that every organic

being, if subjected during several generations to new and varying conditions, would vary. . . . "[79]

Darwin is very certain on the issue of limits to variation, however: "That a limit to variation does exist in nature is assumed by most authors, though I am unable to discover a single fact on which this belief is grounded";[80] ". . . it was shown that no ascertained limit to the amount of variation is known. . . . "[81]

Darwin felt that a key problem in understanding the existence of variations in nature was that they were not readily visible. They were, as he put it, "quite wanting (as far as our senses serve). . . . "[82] This was a problem similar to the lack of intermediate forms—there was no visible evidence. As in the notebooks, the problem of intermediate forms receives considerable attention in the *Essays*.

"This want of evidence of the past existence of almost infinitely numerous intermediate forms, is, I conceive, much the weightiest difficulty on the theory of common descent . . . ," Darwin wrote;[83] ". . . the imperfect evidence of the continuousness of the organic series, which, we shall immediately see, is required on our theory, is against it; and is the most weighty objection. . . . "[84] "I believe it [reasons for rejecting the theory of common descent] is because we are always slow in admitting any great change of which we do not see the intermediate steps," Darwin concluded.[85] The problem of intermediate forms also applied to the question of instincts. "We are not justified in *prima facie* rejecting a theory of the common descent of allied organisms from the difficulty of imagining the transitional stages in the various now most complicated and wonderful instincts," Darwin argued.[86]

One senses from the *Essays*—though the notebooks suggest it too—that Darwin feels that one of the key problems facing his theory is common sense, believability: people cannot see variations in nature or intermediate forms and hence reject his ideas on those subjects. Darwin wrote, for example, about the variety of organic beings in nature: "Our first impression is to disbelieve that any secondary law could produce infinitely numerous organic beings. . . . "[87] He admitted that the development of an organ as complex as the eye " . . . at first accords better with our faculties to suppose that . . . [it] required the fiat of a Creator. . . . "[88]

This problem of common sense which confronted Darwin's theory put him in an awkward argumentative position. On the one hand, his main argument against the Creationist view was based on reasonableness, common sense, and logic: the Creationist position does not really explain phenomena; it does not help us understand the facts of nature; it often simply goes against common sense. On the other hand, Darwin apparently knew that the "common sense" argument could just as easily be turned against his theory: we do not see intermediate forms; common sense tells us there are demonstrable limits to variability. So the common sense argument became for Darwin a very dangerous two-edged sword.

Darwin is also concerned in the *Essays,* as before, with the strategy of his argument, how best to argue his case, and once again he advises himself on what to say and how best to say it. He wrote these instructions in the margins of the manuscript pages and sometimes on the backs of pages; they were recorded by Francis Darwin.

Darwin writes about multiple centers of creation, for example, "Better begin with this. If species really, after catastrophes, created in showers over world, my theory false."[89] "Discuss one or more centres of creation: allude strongly to facilities of dispersal and amount of geological change: allude to mountain-summits afterwards to be referred to."[90] About the development by natural selection of complex organs, Darwin advises: "Certainly (two pages in the MS.) ought to be here introduced, viz. difficulty in forming such organ, as eye, by natural selection."[91] Regarding instincts: "Give some definition of instinct, or at least give chief attributes . . . ,"[92] and regarding species: " . . . here discuss *what is a species,* sterility can most rarely be told when crossed. — Descent from common stock."[93]

Darwin also provided a series of more general instructions dealing with his overall strategy of argumentation. In order to combat opposition to his ideas, Darwin thought he "ought to state the opinion of the immutability of species and the creation by so many separate acts of will of the Creator" and then knock that opinion down;[94] Darwin felt that in applying his theory to organisms more and more geographically remote, it would be best "to introduce [his theory], saying reasons hereafter to be given, how far I extend theory, say to all mammalia — reasons growing weaker and weaker";[95] as for the development of one species into another, Darwin felt he ought to "give only rule: chain of intermediate

forms, and *analogy;* this important. Every Naturalist at first when
he gets hold of new variable type is *quite puzzled* to know what
to think species and what variations";[96] and he decided that the
best arrangement of his argument would be to "give sketch of
the Past, — beginning with facts appearing hostile under present
knowledge, — then proceed to geograph. distribution, — order of
appearance, — affinities, — morphology &c &c."[97]

Darwin's desire to deal candidly with problems confronting his
theory continues to be emphasized. In the notebooks this desire
had amounted to a strong intention; in the *Essays* it is a decided
fact. In the fourth chapter of the 1844 *Essay,* for example, Darwin mentions "the imperfect evidence of the continuousness of
the organic series, which . . . is required on our theory, . . . [and
which] is against it; and is the most weighty objection."[98] He
entitles the third chapter of the 1844 *Essay* "On the Variation of
Instincts and Other Mental Attributes under Domestication and
in a State of Nature; On the Difficulties in this Subject; and on
Analogous Difficulties with Respect to Corporeal Structures."[99]

Darwin is still concerned in the *Essays* with distinguishing his
theory from those of others. He is especially concerned again with
the Creationist view. His strategy of attack against that view is,
as before, to question its usefulness while suggesting its seeming
improbability. " . . . It becomes highly improbable that they [or-
ganisms] have been separately created by individual acts of the
will of a Creator . . . ," Darwin writes.[100] "Shall we then allow
that the three distinct species of rhinoceros which separately inhabit Java and Sumatra and the neighboring mainland of Malacca
were created, male and female, out of the inorganic materials of
these countries? . . . " " . . . Shall we say that without any apparent cause they were created on the same generic type with the
ancient woolly rhinoceros of Siberia? . . . "[101] He concludes: "It
is impossible to reason concerning the will of the Creator, and
therefore, according to this view, we can see no cause why or why
not the individual organism should have been created on any fixed
scheme."[102]

Darwin is concerned, as before, with the law-like basis of the
natural world, and frequently his criticisms of the Creationist
position are closely linked to this emphasis. He contends that it
"accords with what we know of the law impressed on matter by
the Creator, that the creation and extinction of forms, like the

birth and death of individuals should be the effect of secondary [laws] means."[103] He thinks it is folly to believe "that the planets move in their courses, and that a stone falls to the ground, not through the intervention of the secondary and appointed law of gravity, but from the direct volition of the Creator."[104]

Finally, there is an interesting new emphasis in the *Essays* on a general ignorance of the facts of nature, and Darwin makes a connection between that ignorance and the lack of appreciation for the value of his theory. He maintains that the difficulties that one encounters with his theory "are those which would naturally result from our acknowledged ignorance" of the facts of nature.[105] We "were justified in assuming individual creations," Darwin believes, only "as long as species were thought to be divided and defined by an impassable barrier of *sterility*" and "whilst we were [still] ignorant of geology."[106] The reason one might be skeptical of his ideas "is due to ignorance necessarily resulting from the imperfection of all geological records."[107]

The Long Manuscript

Pressured by Lyell in 1856 to publish at least something on his theory—lest, as Lyell prophetically warned, Darwin lose priority for his ideas—Darwin began a short sketch with publication in mind. Given Darwin's penchant for detail and his growing belief that a short sketch would be inadequate, between 1856 and 1858 this sketch became a lengthy, though incomplete, essay on species. I will refer to this essay as Darwin's Long Manuscript.[108]

I do not wish to go into detail about the Long Manuscript, primarily because it is very similar to the *Origin,* which was consciously drawn as an abstract of it; I will be discussing the *Origin* in some detail in the next chapter. But I do want to note briefly some aspects of it which are important.[109]

Darwin seems especially concerned in the Long Manuscript with possible common-sense objections that might be raised against his theory. He feels, for example, that because we rarely see destruction or struggle among domestic animals, the domestic setting is often a distortion of what really happens in nature.[110] He emphasizes the need to reflect mightily on the struggle for existence in nature in order to realize its full extent and pervasiveness.[111]

Darwin also seems concerned, as he had been in the *Essays*, with the believability of his theory in light of the general ignorance of so many of the facts of nature. Many of the difficulties in understanding his theories are due to such ignorance: " . . . our great ignorance of the complete biography of any one single plant or animal makes us slow to believe in the multiform & often extremely obscure checks to their increase."[112] The laws of variation are also unfamiliar to us;[113] regarding a classification system based on hereditary characters, Darwin observes "that we are far too ignorant to apply it to varieties under natural conditions, more especially in regard to animals."[114]

Darwin is determined in the Long Manuscript to fulfill his earlier intention of discussing all problems fully and openly. He has, for example, a section entitled "Facts apparently opposed to there being a severe struggle in all nature." Dealing openly with apparent instances of nonstruggle or where injured organisms have survived,[115] Darwin gives cases "which alone have seemed to me to throw doubt on the struggle for existence."[116]

Finally, he emphasizes two new themes in the Long Manuscript which we will see often in the *Origin*. One is the imposed brevity which Darwin feels prohibits him from providing the fullest and strongest arguments possible. Robert Stauffer, editor of Darwin's Long Manuscript, maintains that "even the . . . [long] manuscript had been for Darwin a condensed form of the presentation he preferred for his material. . . ."[117] There is ample evidence to support Stauffer's contention. As Darwin wrote uncomfortably to Lyell on November 10, 1856 concerning his work on the Long Manuscript: "I have found it quite impossible to publish any preliminary essay or sketch; but am doing my work as completely as my present materials allow without waiting to perfect them."[118] To Hooker on December 10, 1856 a frustrated Darwin wrote: "It is a most tiresome drawback to my satisfaction in writing that, though I leave out a good deal and try to condense, every chapter runs to such an inordinate length."[119] Darwin begins a section in the Long Manuscript on "Checks to increase in animals" by stating: "A volume would be required to treat the subject properly, & I can give here only a few of the leading facts, which have most struck me."[120]

The other theme concerns the complexity of the relationships among all organisms in nature. Because struggle among organisms is "often [of] a very complex nature," Darwin finds "many points of present inextricable confusion" surrounding a theory of the descent of species.[121] This theme is closely related to Darwin's emphasis on ignorance, and serves a similar explanatory function: his theory cannot be fully appreciated because nature is too complex to be easily understood in the terms in which Darwin is analyzing it.

The Origin

On June 18, 1858, while working on his Long Manuscript, Darwin received a letter from Alfred Russel Wallace, enclosing a description of a theory of evolution almost identical to Darwin's. Some of the details of what followed will be examined in Chapter 8. In short, Darwin immediately began to prepare an abstract of his Long Manuscript for publication. He finished the abstract in eighteen months, and in November 1859 the *Origin* was published.

There are several important differences between the Long Manuscript and the *Origin*. First, Darwin does not pursue the open discussion of problems to the same degree in the *Origin*, even though he is still determined to confront them candidly. For example, in the *Origin*, probably because of lack of space, Darwin does not probe deeply the possibility that there may in fact be no struggle for existence; whereas in the Long Manuscript he considers the counter-evidence at length.

Also, the *Origin* was edited for publication and presentation to the public; the Long Manuscript was not. I think that, as a result, the *Origin* represents a later, more refined, and clearer statement of Darwin's theories. Darwin felt so too. In the letter written to Hooker dated October 6, 1858, we will remember, Darwin revealed that writing the *Origin* as an abstract of his Long Manuscript "has clarified my brains very much, by making me weigh the relative importance of the several elements."[122] Finally, I agree with Stauffer's contention that the Long Manuscript has more examples illustrating Darwin's various ideas than the *Origin*. It also has extensive footnotes, which the *Origin* does not. Although the Long Manuscript is more scholarly than the *Origin* and treats subjects at far greater length, it is much less direct in its argument, and much less clearly focused.

Let us now look at the argument of the *Origin* in detail.

7

The *Origin* as the Least Objectionable Theory

Presenting the Argument

Organization of the Origin

The *Origin* seems to fall quite naturally into five major parts: an introduction; presentation of Darwin's theory (Chapters I through V); a discussion of major objections to the theory (Chapters VI through IX); evidence adduced in support of the theory (Chapters X through XIII); and a final summary and conclusion (Chapter XIV).

Darwin begins the *Origin* very cautiously, half apologizing in the introduction for publishing the present work.[1] He labels the *Origin* an abstract of a longer, intended work—an abstract which, he warns the reader, "must necessarily be imperfect." He will not be able to give references and authorities for his statements. He also wants the reader to know that the same facts with which he will support his own theory can support opposing theories as well, and that he is "well aware that scarcely a single point is discussed in this volume on which facts cannot be adduced, often apparently leading to conclusions directly opposite to those at which I have arrived."[2]

At the same time, Darwin is setting the stage for the *Origin*, not as a dispassionate analysis of the question of the descent of species, but as a vigorous argument in support of his own theory. He has abandoned his long-held intention of presenting both sides of the question. He still feels that a "fair result can be obtained only by fully stating and balancing the facts and arguments on both sides," but unfortunately, under present limitations of space, "this cannot possibly be . . . done."[3]

119

Darwin's decision to argue only for descent may have been an additional source of anxiety for him. He could no longer, in effect, take refuge behind an "objective" analysis of both sides of the issue, but must present his arguments openly before the public.

Darwin suggests that the key question is not whether species have descended from other species, but rather how that process might have occurred.[4] There are certain facts that Darwin feels he needs to establish. He hopes to show that hereditary modification is at least possible by proving that variations occur and that they are sometimes inherited. Also, he wants to show that man, in a relatively short period of time, has been able to effect great changes in domestic species. He needs to do this in order to set up the possibility of natural selection working in nature. Darwin stresses that in order to understand the sorts of changes that can occur in a domestic setting, one must look at the large body of work on domestic breeding—work which he feels has "been very commonly neglected by naturalists."[5] Domestic breeding becomes the primary new source of information presented in the *Origin* and the most important analogy Darwin uses for purposes of analyzing what could happen in nature. Much remains to be known about the origin of species, Darwin concludes. The view that species had been independently created—"the view which most naturalists entertain, and which I formerly entertained,"—is erroneous. Natural selection is the principal, but by no means the sole, source of change in the organic world.[6]

Explicating the Theory

Chapter I, "Variation Under Domestication," begins a five-chapter explication of Darwin's theory. The major point that Darwin wants to establish in this first chapter is that in a domestic setting variations occur and are inherited. He has to establish the credibility of his domestic breeding analogy. Darwin thinks it obvious that variations occur—that is the purpose of domestic breeding. Darwin points out that no breeder doubts the strong tendency to inheritance among domestic breeds—"doubts have been thrown on this principle by theoretical writers alone."[7] He believes that the inheritance of variation is the rule, and noninheritance the anomaly.[8] The occurrence of variation and inheritance leads to the great power of artificial selection, and Darwin points to several breeds of cattle and sheep that man has been able to change in a relatively few generations, "even within a single lifetime."[9]

On the causes of variation, Darwin theorizes that "variability may be largely attributed to the ovules or pollen . . . having been affected by the treatment of the parent prior to the act of conception."[10] He contends that compared to the laws of reproduction, growth, and inheritance, the direct effects of the conditions of life are unimportant as a cause of variation, though he thinks that use and disuse may have had some influence.[11] Also, Darwin rejects the contemporary idea of reversion—that is, the idea that domestic species, when allowed to run wild, revert to their aboriginal stocks.[12] It is important for Darwin to reject this idea; if reversion did occur, he would be unable to draw meaningful deductions from domestic species to species in a state of nature.

Darwin emphasizes three ideas or themes throughout this chapter. One, which was discussed in the Long Manuscript, is that nature is frighteningly complex. For example, " . . . the result of the various, quite unknown, or dimly seen laws of variation is infinitely complex and diversified . . . ";[13] the final result of the various causes of variation "is . . . rendered infinitely complex."[14] Another theme, first emphasized in the *Essays*, is that we are ignorant of the causes of many phenomena. Darwin believes, for example, that our understanding of the species from which domestic productions are descended "must . . . remain vague"; and that " . . . the laws governing inheritance are quite unknown. . . . "[15] Finally, the theme of the abstract nature of the *Origin*—a theme first developed in the Long Manuscript—is emphasized again. Darwin feels that while he has discussed in detail the probable origin of domestic pigeons, for example, he has done so at insufficient length.[16]

The primary purpose of Chapter II, "Variation Under Nature," is to show that variations, which Darwin has shown to exist in a domestic setting, exist in nature as well. He does this indirectly through a long discussion of the differences between species and varieties.[17] Darwin stresses the difficulty of distinguishing between the two: " . . . few well-marked and well-known varieties can be named which have not been ranked as species by at least some competent judges."[18] Darwin notes at another point: "I was much struck how entirely vague and arbitrary is the distinction between species and varieties."[19]

To Darwin, varieties are merely incipient species, which means, in effect, that there is constant variation occurring in nature. The problem is that naturalists have not really focused very carefully on these minute changes.[20] Darwin views the term "species" as merely a term of convenience, not having any particular substantive meaning in the natural world or describing organisms that are very different from those described by the term *variety*.[21] Darwin suggests that the development of species into subspecies and then into new species must be attributed to the work of natural selection accumulating slight variations in a definite direction. There is no mention yet of how this selection process might work, nor is the analogy with domestic, artificial selection drawn.

A new important theme emerges in this chapter of the *Origin*, though it was emphasized before in the Transmutation and M and N notebooks, the 1842 and 1844 *Essays*, and the Long Manuscript. This theme is Darwin's favorable contrast of his theory with the Creationist point of view: with the theory of natural selection, certain phenomena in nature make sense, while from a Creationist point of view they make no sense at all. The relationships Darwin describes in this chapter between species and varieties, between large genera and their numerous specific groups, make sense, Darwin argues, if we assume that "species have once existed as varieties . . . whereas, these analogies are utterly inexplicable if each species has been independently created."[22]

Finally, Darwin again emphasizes the problem of lack of space. To treat the subject of variation in nature in any detail, Darwin would have to give "a long catalogue of dry facts" which, because of space limitations, he will have to reserve for a longer, future work.[23]

In Chapter III, "Struggle for Existence," Darwin suggests that merely the existence of variation in nature, though important, helps us little in understanding how species arise in nature.[24] The key question is how are new species developed and adapted. Darwin suggests that all adaptation and change in nature occur through an intense struggle for existence among organisms. Darwin mentions De Candolle's and Lyell's descriptions of struggle, but he emphasizes that his view of struggle recognizes an even more intense battle. Darwin sees Malthus's doctrine of the imbalance between population growth and food supply applied with even greater force to nature: organisms will tend to increase geometrically while the food supply increases only arithmetically; as a consequence, there will be a severe struggle among organisms for a limited number of places in nature; much destruction will occur; organisms having some advantage will survive.[25]

Darwin emphasizes that he is using the term *struggle for existence* in a large and metaphorical sense, including successful reproduction. Darwin's figurative struggle, defined as dependency between organisms, usually occurs between species remote in the scale of nature; literal struggle is intense between organisms within the same or closely related species.[26]

Two themes are emphasized throughout this chapter. One is that the conditions of life, which Darwin describes as the competitive relationships among organisms, and not external conditions such as climate and geography,[27] are key to changes in organic structure. A good example is that the arrival of a new competitor in an area might severely change a given organism's conditions of life, though the climate remains the same. The other idea is again our profound ignorance, this time regarding the struggle for existence in nature and the complex relations of all organic beings. We simply have not been able to appreciate how severe such struggle is,[28] and naturalists have not really focused much on the complex relations among organisms.[29]

Chapter IV, "Natural Selection," refers to topics discussed in Chapter I and attempts to determine if the principle of selection, so potent in man's hands, can be applied to nature with equal force. Darwin emphasizes once again how much variation there is in nature, and how strong the hereditary tendency is. He also stresses that organisms having some advantage in the struggle for existence will survive. Unless profitable variations do occur, natural selection can do nothing.

Darwin cites Lyell's *Principles of Geology* as a point of reference for the cumulative effect over time of minor changes.[30] As Lyell's new geology has banished such views as the excavation of a great valley by a single diluvial wave, "so will natural selection," Darwin writes, "if it be a true principle, banish the belief of the continued creation of new organic beings, or of any great and sudden modification in their structure."[31]

Darwin discusses the concept of divergence of character in an effort to understand how species and varieties eventually differ. He uses the analogy of domestic breeding to shed light on this question. The key to divergence, Darwin feels, is that the more diversified an organism becomes, the better that organism will be able to survive in a changing environment. Better adapted organisms will eventually outdistance their progenitors, which eventually become extinct. This is why we often find extinct species closely linked geographically to similar living forms.[32] Darwin presents a long, hypothetical example of how the natural selection process might work in nature and how divergence of character might occur.[33] In this discussion, he touches briefly on a number of topics—the intercrossing of species, correlation of growth, natural classification, sexual selection—some of which he discusses in greater detail in later chapters.

It is interesting to note that by the end of this chapter, Darwin has begun to refer to several of the elements of his theory of descent with modification by natural selection as "indisputable"— elements which a few pages before Darwin was attempting to substantiate as actually occurring in nature. He maintains that organisms vary ("this cannot be disputed"[34]) and that there is a severe struggle for existence among organisms in nature ("this certainly cannot be disputed"[35]). Darwin contends that his theory, whether true or not, "must be judged of by the general tenour and balance of evidence given in the following chapters."[36]

Several of the familiar themes of previous chapters appear again in Chapter IV. Again Darwin feels that he has not been able to discuss his evidence in sufficient detail. He admits that his discussion of the struggle for existence in the last chapter, for example, was too brief.[37] He also refers to the subject of the intercrossing of species: "I must here treat the subject with extreme brevity . . . as it is impossible . . . to enter on details. . . ."[38] He mentions as well that he has collected "many special facts" regarding intercrossings "but . . . I am not here able to give [them]."[39]

Darwin once again emphasizes the complexity of the ideas and phenomena with which he is dealing. He indicates that the circumstances favorable to natural selection comprise "an extremely intricate subject."[40] He also mentions the "extreme intricacy" of circumstances favorable and unfavorable to natural selection,[41] describes divergence of character as "this rather perplexing subject,"[42] and notes again the "infinite complexity" of the relations of organisms in nature.[43]

Finally, he reemphasizes the theme of natural selection's explanatory superiority over the other competing theories, primarily the Creationist view. The intercrossing of species is a law of nature in which, Darwin writes, "we can, I think, understand several large classes of facts . . . which on any other view are inexplicable."[44] Darwin feels his theory helps explain a system of natural classification which "on the view that each species has been independently created, I can see no explanation for. . . ."[45]

Chapter V, "Laws of Variation," instead of a chapter by which, as Darwin had suggested, the truth of his theory may be judged, is, in fact, a return to the question of the causes of variation briefly discussed in Chapter I, and is largely a restatement of that discussion and of other topics previously mentioned. The major point of Chapter V is that we know very little about the origins of variations. Darwin suspects that variations are not due to chance, but he is not certain. We are forced, Darwin suggests, "to acknowledge plainly our ignorance of the cause of each particular variation."[46] We are, in Darwin's words, "profoundly ignorant" of why one part or another of an organism should vary more or less than any other.[47]

Our profound ignorance of the laws of variation extends also to some aspects of the process of acclimatization (the acclimation of organisms to new or changed climates), the causes of which remain "a very obscure question."[48] Darwin describes correlation of growth (the idea that, because organisms are so closely tied together internally, when any one part is modified, there is a tendency for closely correlated parts to be modified as well) as "a very important subject, most imperfectly understood," while "the nature of the bond of correlation is frequently quite obscure" as well.[49] We are, in addition, ignorant of why in some organisms we see a reversion to lost ancestral characteristics. Darwin here is referring specifically to the odd appearance of stripes on some horses.[50]

Darwin regrets that because of space limitations he cannot provide a long list of facts to support his idea that variations are probably caused by the susceptibility of the reproductive system to changes in the conditions of life.[51] Also, he regrets not being able to provide a long list of examples of variations in organs, "but here, as before," he writes, "I lie under a great disadvantage. . . ."[52]

Finally, he emphasizes again his ability to explain certain phenomena which from the Creationist view make no sense at all. Regarding the variability of rudimentary organs, Darwin writes: "On the view that each species had been independently created, with all its parts as we now see them, I can see no explanation. But on the view that groups of species have descended from other species, and have been modified through natural selection, I think we can obtain some light."[53] Darwin makes a similar point about the high variability of abnormally developed characteristics.[54]

Difficulties on the Theory

Up to this point, Darwin has accomplished the following: he has postulated his theory of descent with modification by means of natural selection with several of its "indisputable" component parts, and has at least raised the possibility that his theory might be true; he has attempted to support his theory by stressing what it can help explain and by contrasting it favorably with other theories attempting to explain the same phenomena—primarily the Creationist view; he has painted a landscape of ignorance and doubt, uncertainty and complexity regarding the problems involved in attempting to understand the complex relationships of organisms in nature and the often obscure processes by which these organisms might have evolved, with the implication that people might doubt the validity of his theory because of such ignorance and complexity; and he has attempted to excuse the frequently attenuated nature of many of his discussions by pointing to the severe space limitations under which he is working.

In Chapter VI, "Difficulties on the Theory," Darwin begins a major discussion of difficulties surrounding his theory, conducting it, not unexpectedly, with great candor. He suggests that by this time many difficulties must have become apparent to the reader. "Some of them are so grave that to this day I can never reflect on them without being staggered . . . ," Darwin admits, "but, to the best of my judgment, the greater number are only apparent, and those that are real are not, I think, fatal to my theory."[55] He then lists the key difficulties: the first is the familiar question of intermediate forms. If species have descended from others by insensibly fine gradations, why do we not see everywhere innumerable transitional forms? The second concerns the seemingly contradictory achievements of nature. Would or could nature produce both trifling organs or characteristics (such as the tail of a giraffe) and also the eye, an organ of great complexity and perfection? The third concerns instincts, a problem, as we have seen, that Darwin has been struggling with since the Transmutation Notebooks. Can instincts be acquired and modified through natural selection? even such as the marvelous cell-making instincts of the hive-bee? The fourth concerns the question of sterility. How can we account for sterility when species cross yet fertility when varieties cross? Are not species specially endowed with the quality of sterility in order to prevent the confusion in nature of all organic forms? Darwin discusses the first two difficulties in Chapter VI, the last two in Chapters VII and VIII, respectively.

Darwin feels that the absence of transitional forms can be explained by the fact that the natural selection process itself tends to exterminate intermediate forms, and that intermediate forms usually endure for only a short period of time.[56]

Darwin admits, with respect to organs of extreme perfection and complication, that the idea that the eye was formed by natural selection "seems, I freely confess, absurd in the highest possible degree."[57] But he suggests that the eye, even in its transitional form, was most probably of some advantage and use to the organism. In any case, he feels that such a difficulty as this should not in itself cause one to reject his theory. "He who will go this far, if he find on finishing this treatise that large bodies of facts, otherwise inexplicable, can be explained by the theory of descent," Darwin argues, "ought not to hesitate to go further, and to admit that a structure even as perfect as the eye of an eagle might be formed by natural selection, although in this case he does not know of the transitional grades. His reason ought to conquer his imagination; though I have felt the difficulty far too keenly to be surprised at any degree of hesitation in extending the principle of natural selection to such startling heights."[58]

Darwin suggests that he has as great a problem understanding the origins of organs of little apparent importance as he does with complicated organs like the eye. But "we are much too ignorant in regard to the whole economy of any one organic being," he writes, "to say what slight modifications would be of importance or not."[59] Organs of little importance now might have been of considerable importance millenia ago, and these organs might merely have been carried through to successive generations by heredity.[60]

"[There are] . . . difficulties and objections which may be urged against my theory," Darwin writes in summarizing Chapter VI. "Many of them are very grave; but I think that in the discussion light has been thrown on several facts, which on the theory of independent acts of creation are utterly obscure."[61] Darwin further maintains that although the idea that an organ so perfect as the eye was formed by natural selection "is more than enough to stagger anyone," there is yet "no logical impossibility in the acquirement of any conceivable degree of perfection through natural selection."[62]

In Chapter VI Darwin emphasizes again the lack of space which has seriously handicapped his discussions. He writes about transitional forms, for example: "Here, as on other occasions, I lie under a heavy disadvantage, for out of the many striking cases which I have collected, I can give only one or two instances. . . . "[63]

The idea of the superiority of natural selection over the Creationist view is prominent again as well. "He who believes that each being has been created as we now see it, must occasionally have felt surprise when he has met with an animal having habits and structure not at all in agreement. . . . ," Darwin writes.[64] An example of this is the webbed feet of upland geese, who almost never go near the water.

A new theme emerges in this chapter and remains prominent throughout the remainder of the *Origin:* the fatal objection that could destroy Darwin's theory. At one point in his discussion, Darwin suggests that many naturalists recently had raised a protest against the utilitarian idea that every detail of structure has been produced for the good of its possessor. "They believe that very many structures have been created for beauty in the eyes of man, or for mere variety . . . ," Darwin writes. "This doctrine, if true, would be absolutely fatal to my theory."[65] Yet it is not true; natural selection shows that every detail of structure is important to the organism or was important to some ancestor; natural selection will always work for the good of the organism, and it will never produce an organ injurious or initially of no use to its possessor.

In Chapter VII, "Instincts," Darwin takes up the third major difficulty he sees standing in the way of acceptance of his theory. How can instincts be the product of natural selection, Darwin asks, especially those as complex as the cell-making instinct of the hive-bee? Darwin suggests that this particular instinct is "a difficulty sufficient to overthrow my whole theory,"[66] though he thinks that it can be explained by the slow accumulation, through natural selection, of the hive-bee's inimitable architectural powers.[67] Other peculiar instincts can be explained in a similar manner.[68]

Darwin candidly admits that he cannot claim that "the facts given in this chapter strengthen in any great degree my theory; but none of the cases of difficulty, to the best of my judgment, annihilate it."[69] On a more positive side, Darwin argues, the doctrine of *"natura non facit saltum,"* nature makes no leaps, applies to instincts as well as to bodily parts, with the implication that instincts, just like bodily characteristics, can be the result of the slow accumulation of successive, minute modifications.[70]

In Chapter VIII, "Hybridism," Darwin discusses the last of the major difficulties outlined earlier. His main point here is that there are no general laws of breeding prohibiting the development of new species. Species are not always sterile when they are crossed, and varieties are not always fertile when they are crossed. Darwin suggests, in fact, that some intercrossing often strengthens the breed and is, therefore, an aid to various groups of organisms in the struggle for existence.[71]

Darwin maintains that sterility is not a specially acquired or endowed quality, but is often incidental to other acquired differences, and its causes are unknown. The idea that species have been specially endowed with the quality of sterility in order to prevent their crossing and blending together in utter confusion—a notion which Darwin attributes to the Creationists—is severely criticized. "[Have] . . . species . . . been endowed with sterility simply to prevent their becoming confounded in nature?" Darwin asks. "I think not. For why should the sterility be so extremely different in degree, when various species are crossed, all of which we must suppose it would be equally important to keep from blending together?"[72]

In Chapter VIII, Darwin once again emphasizes his disadvantage in not being able to discuss topics at greater length. He prefaces a discussion of a comparison of hybrids and mongrels with the caution: "I shall here discuss this subject with extreme brevity."[73] At another point he mentions facts "briefly given in this chapter. . . ."[74]

Fossil Remains and the Problem of Transitional Forms

In Chapters VI, VII, and VIII of the *Origin* Darwin discussed, as we have seen, the chief objections which could be waged against his theory. In Chapter IX, "On the Imperfection of the Geological Record," he returns to the first problem discussed—the lack of existing transitional forms. Darwin reiterates the main reasons why he thinks this "very obvious difficulty" is not "fatal" to his views.[75] That still leaves the question, he admits, of the fossil remains of transitional forms,[76] the lack of which he terms "the most obvious and gravest objection which can be urged against my theory."[77] But he is not willing to give up in the face of this problem. He feels that the explanation for the lack of fossil remains lies "in the extreme imperfection of the geological record."[78]

Using Lyell's *Principles of Geology* again as a point of reference, Darwin emphasizes the enormous period of time that has elapsed since organic life first appeared, and he stresses the paltry display in museums of the remains of all the creatures that must have existed at some time. Also, the record of the fossil remains that we do have is unrepresentative and imperfect: soft shelled organisms simply will not be preserved; hard shelled ones will be preserved only under rare circumstances. As a result, "we have no right," Darwin argues, "to expect to find in our geological formations, an infinite number of those fine transitional forms, which on my theory assuredly have connected all the past and present species of the same group into one long and branching chain of life."[79]

Darwin summarizes Chapter IX:

The several difficulties here discussed . . . are all undoubtedly of the gravest nature. We see this in the plainest manner by the fact that all the most eminent palaeontologists, namely Cuvier, Owen, Agassiz, Barrande, Falconer, E. Forbes, &c., and all our greatest geologists, as Lyell, Murchison, Sedgwick, &c., have unanimously, often vehemently, maintained the immutability of species. . . . For my part, following out Lyell's metaphor, I look at the natural geological record, as a history of the world imperfectly kept, and written in a changing dialect; of this history we possess the last volume alone, relating only to two or three countries. Of this volume, only here and there a short chapter has been preserved; and of each page, only here and there a few lines. . . . On this view, the difficulties above discussed are greatly diminished, or even disappear.[80]

Evidence in Support of the Theory

Having presented his theory and surveyed the chief problems that might affect acceptance of his ideas, Darwin turns in Chapters X through XIII to the other side of the question: obvious evidence in support of his theory.

Chapter X, "On the Geological Succession of Organic Beings," is, in some sense, a companion to Chapter IX; whereas in Chapter IX Darwin defends his theory against a great mass of paleontological evidence that does not exist, in Chapter X he attempts to support his theory by reference to paleontological evidence that does exist. His method of argument is by now a familiar one to the reader: his theory can explain certain phenomena which with any other theory, particularly the Creationist view, simply do not make much sense.

Darwin writes, for example, about the fact that the forms of life seem to change almost simultaneously throughout the geologic world: "This great fact of the parallel succession of the forms of life throughout the world, is explicable on the theory of natural selection."[81] This is because new forms that might get an advantage over others would tend to destroy existing forms and new species would be formed by dominant species varying and spreading widely.[82] From the Creationist point of view, this fact makes little, if any, sense. Darwin also feels that the affinities of extinct and living species "all fall into one grand natural system"; that this fact "is at once explained on the principle of descent," based, as it is, on a genealogical view of their relationship; while no explanation exists from a Creationist point of view.[83]

In Chapters XI and XII, both on "Geographical Distribution," Darwin continues to argue natural selection versus Creationism: the present geographical distribution of organisms throughout the world makes sense in terms of the combined effects of natural selection and migration emanating from a single center of creation, but makes little sense from a Creationist point of view.

In Chapter XI, Darwin points out that various physical and climatic conditions, when we look at the geographical distribution of organisms throughout the world, have probably had little role to play in the eventual form of these organisms. We see great physical similarities between the Old and New worlds, for example,

but the organisms that inhabit each area are usually quite different.[84] Family and genealogy are much more significant links between organic forms than similarity of climate or geography.

On the question of whether species have been created at one
or more points on the earth's surface, a question he struggled with
in the notebooks and the *Essays*, Darwin writes: "If the same species can be produced at two separate points, why do we not find
a single mammal common to Europe and Australia or South
America? . . . The answer, as I believe, is, that mammals have
not been able to migrate, whereas some plants . . . have migrated
across the vast and broken interspace."[85] Darwin believes that
species have been produced in one area and from there have migrated. He quotes A. R. Wallace to the effect that every species
coming into existence does so coincident both in space and time
with a preexisting, closely allied species, thus suggesting their
familial origins.[86]

Darwin then turns to a lengthy discussion of the means of geographical dispersal and migration.[87] While Lyell and others have
treated the subject in detail, here he "can give . . . only the briefest abstract of the more important facts."[88] He presents some results of his researches and ideas on means of dispersal, including
his salt-water seed experiments, his theories about transport of
seeds by birds, and his ideas on transport by iceberg—all of which
focus on the idea that the present geographical distribution of organisms throughout the world can be explained in terms of present laws of migration and dispersal without reference to special
acts of creation.

Regarding the sticky problem of how organisms in high and
distant mountain areas could so clearly resemble each other without any seemingly reasonable means of migration between them,
Darwin offers his theory of the geographical distribution of organisms during the glacial period. He believes that during the great
glacial epochs, areas now widely separated, such as distant mountain peaks, were connected by glaciers. When warmer weather
returned and the glaciers began to recede, the geographical areas
once again became isolated from one another.[89] Darwin admits
that "very many difficulties remain to be solved" on this theory
and that a lot of these issues "cannot be . . . discussed" in detail
in his present work.[90]

In Chapter XII, Darwin discusses the geographical distribution of organisms on oceanic islands. He asks why, according to the theory of Creation, should not whole groups of organisms appear on these islands? The absence of such groups "would be very difficult to explain"[91] from the Creationist viewpoint, Darwin maintains. Natural selection, however, explains this absence by the fact that the geographical distance between mainland areas and oceanic islands has prevented whole groups of organisms from migrating.[92]

Darwin points to the affinity of oceanic species with those of the nearest continent or mainland as "the most striking and important fact" in support of his theory;[93] he describes the relation among organisms in the Galapagos Islands as an example of the affinity between island and continental productions of the same area.[94]

Darwin summarizes the two chapters on geographical distribution, emphasizing the new understanding that the theory of natural selection brings to the subject: "If the difficulties be not insuperable in admitting that in the long course of time the individuals of the same species, and likewise of allied species, have proceeded from some one source; then I think all the grand leading facts of geographical distribution are explicable on the theory of migration (generally of the more dominant forms of life), together with subsequent modification and the multiplication of new forms."[95]

In Chapter XIII, "Natural Affinities of Organic Beings; Morphology; Embryology; Rudimentary Organs," Darwin adduces more evidence in support of his theory, strongly contrasting once again its explanatory powers with those of the Creationist view. He feels that all organisms fall into a natural classification system, the inevitable result of descent with modification.[96] Darwin suggests that morphology teaches us that there is a close structural resemblance among all organisms: the hand of a man, the hand of a mole, the leg of a horse, the paddle of a porpoise, the wing of a bat—all are constructed on the same pattern, include the same bones, and have these bones in the same relative positions. From the viewpoint of natural selection, these facts of morphology make sense; all organisms are constructed on the same model, having descended from the same prototype.[97] Regarding embryological development, embryos are often so much alike that it is difficult to know if an embryo is that of a mammal, bird, or reptile.[98] Also, certain organs in an individual, which, when developed, serve different purposes and look dissimilar, are alike in the embryonic stage. How can we explain these similarities? "I believe that all these facts can be explained . . . on the view of descent with modification," Darwin writes.[99] He views the embryo as "the animal in its less modified state; and in so far it reveals the structure of its progenitor."[100] He sees it as "a picture, more or less obscured, of the common parent-form of each great class of animals."[101] With regard to rudimentary, atrophied, or aborted organs, Darwin criticizes the view that they have been created "for the sake of symmetry," or in order "to complete the scheme of nature."[102] To Darwin this is no explanation but merely a restatement of fact. "Rudimentary organs may be compared with the letters in a word," he writes, "still retained in the spelling, but become useless in the pronunciation, but which serve as a clue in seeking for its derivation."[103]

Darwin concludes the chapter by suggesting that the several classes of facts which he has been considering in this chapter "seem to me to proclaim so plainly, that the innumerable species, genera, and families of organic beings, with which this world is peopled, have all descended, each within its own class or group, from common parents, and have all been modified in the course of descent, that I should without hesitation adopt this view, even if it were unsupported by other facts or arguments."[104] Darwin obviously considers the evidence from classification, morphology, embryology, and rudimentary organs to be the strongest in support of his theory.

Concluding the Origin

I suspect that by the time Darwin got to Chapter XIV, "Recapitulation and Conclusion," he had a better understanding of his own theory.[105] The organization of this final chapter is tighter than any of the previous ones. The lengthy discussions of previous chapters are gone. Darwin's arguments seem stronger, more to the point. This final chapter is almost a self-contained mini-*Origin*, organized into neat sections on the theory, on the diffi-

culties with the theory, and on the supporting evidence. The lean-
ness and directness of this final chapter also may relate to the
nature of summaries.

In this final chapter, Darwin describes the *Origin* as "one long
argument,"[106] suggesting that much of what he has been doing
throughout the *Origin* is arguing a case. He says he realizes that
"many and grave objections may be advanced against the theory
of descent with modification through natural selection," and he
has "endeavoured to give to them their full force."[107] He then
recapitulates some of the more serious of these.[108] "I have felt
these difficulties far too heavily during many years to doubt their
weight . . . but it deserves especial notice that the more important objections relate to questions on which we are confessedly
ignorant; nor do we know how ignorant we are. . . . Grave as these
several difficulties are, in my judgment they do not overthrow
the theory of descent with modification."[109]

Darwin then turns to "the other side of the argument," meaning, "the special facts and arguments in favor of the theory." Referring to his domestic breeding analogy, he suggests that if man
can select in a domestic setting, there is "no obvious reason why
the principles which have acted so efficiently under domestication should not have acted under nature."[110] After reiterating the
components of his theory of evolution by natural selection—
variation, a struggle for existence, inheritance, divergence of character, and extinction—Darwin suggests that this theory, "even if
we looked no further . . . seems to me to be in itself probable."[111]
To support his point further, he lists natural phenomena that "we
can understand" from a natural selection point of view but which
are inexplicable from a Creationist perspective.[112] The contrast
between his theory and the Creationist view becomes most acute,
most vivid in this final chapter.

Having recapitulated the most convincing facts and consider-
ations that species have changed, and are still slowly changing
by the preservation and accumulation of successive slight favorable variations, Darwin once again asks the pragmatic question:
"Why . . . have all the most eminent living naturalists and geologists rejected this view of the mutability of species?"[113] Darwin
sees two main causes for this rejection: "The belief that species
were immutable productions was almost unavoidable as long as
the history of the world was thought to be of short duration. . . . "

But "the chief cause of our natural unwillingness to admit that one species has given birth to other and distinct species, is that we are always slow in admitting any great change of which we do not see the intermediate steps. . . . The mind cannot possibly grasp the full meaning of the term of a hundred million years. . . ."[114] Darwin notes that the same problem existed when Lyell first presented his revolutionary ideas on geology.

Although he is personally convinced of the truth of his theories, Darwin does not expect to convince many experienced naturalists who have viewed the subject for many years from a completely different point of view. It is obvious that he expected a serious uphill fight for acceptance of his ideas.[115]

In conclusion, Darwin, perhaps in an attempt to wrest God from the side of the Creationists, reiterates the theme he emphasized in his species writings before the *Origin*—the idea that his views accord "better with what we know of the laws impressed on matter by the Creator." Further, since his ideas do not suggest any past, cataclysmic desolation of the world, Darwin assures us that "we may look with some confidence to a secure future of equally inappreciable length."[116] Natural selection, after all, only works towards the good of every being.

The *Origin*'s last analogy is to the mechanistic world of Newton: while this planet "has gone cycling on according to the fixed law of gravity," organisms have evolved, Darwin implies, according to similar fixed laws, suggesting that the laws governing the organic world are really no different in kind from those governing the inorganic.[117]

Problems of Evidence

When we begin to look in detail at the evidence upon which Darwin bases his argument in the *Origin*, we will be better able to appreciate why Darwin, in his letters and in the notebooks, the *Essays*, and the Long Manuscript, often seemed so doubtful and uncertain about the value of his ideas, seemed so concerned about substantive problems surrounding his theory, and felt compelled to think strategically about how best to argue his case and convince people of the value of his ideas.

Before we do that, some of the positive contributions that Darwin was able to make in the *Origin* ought to be stressed, lest we lose sight of the monumental contribution to the history of biology that the *Origin* represents.

There are, of course, many contributions that Darwin made in the *Origin*. In addition to the idea of natural selection, I would mention three. One was his willingness to deal openly with various problems surrounding his theory. This candor made his case for evolution more effective—Darwin felt that it did—even though it helped him little in terms of adding evidence to support his theory.

Another contribution was Darwin's success in removing the discussion of the mutability of species from the realm of religion, mysticism, or faith. In the *Origin* there is no reference to explanations outside the realm of observable or possibly observable phenomena, no mysterious willing of organisms, no teleological tendency to perfection. Where phenomena cannot be explained— the origin of variations, for example—Darwin simply admits his present ignorance.

Finally, and I think most importantly, the *Origin* contains page after page of brilliant argumentation. In the skillfull marshalling of his materials and the deft and often ingenious aligning of his arguments, the depth and quality of Darwin's contribution becomes most apparent. Darwin is able to put together a coherent, cohesive, and forceful argument on a subject of vast magnitude and difficulty supported by a relatively narrow base of solid, incontrovertible evidence.[118]

Now for a look at his evidence. Darwin certainly has strong evidence of extinction. Evidence for other elements of his theory is not as strong. The key supporting evidence of variation, for example, consists of a new and unsubstantiated analogy (domestic breeding),[119] a long discussion of the differences between species and varieties (to show that there are not many, a view which was then certainly arguable), and an assertion of the general myopia of contemporary naturalists when trying to see variations in nature.

In his discussion of the struggle for existence, Darwin mentions the views of several prominent naturalists who were also a bit myopic about the intensity of struggle; he offers a hypothesis based on a theory of how a struggle for existence might occur in nature, using in large part Malthus's theory of human populations. Finally, and more substantially, he adduces evidence from several of his experiments at Down on seeds and young plants.[120] Darwin depends solely for proof of the inheritance of acquired characteristics on an analogy to domestic breeding where man, unlike nature, is able consciously to control breeding patterns and growth. Perhaps the key concept in Darwin's theory on divergence of character —how forms eventually evolve—again uses domestic breeding as an analogy and provides a hypothetical example of how divergence of character might occur in nature.

The most Darwin could hope for from a discussion of his natural selection theory per se was to present the possibility that such a process might have occurred and is presently occurring in nature. He certainly accomplishes this. The evidence that it probably did occur, he feels, would have to come from elsewhere.[121] This may be why Darwin suggests in Chapter IV of the *Origin*, following the presentation of his theory, that the truth of his theory would have to rest on "the general tenour and balance of evidence" in the remaining chapters. He seems to reverse himself, however, in Chapter XIV when, after reiterating his argument, he suggests that "even if we looked further . . . [the theory] seems to me to be in itself probable." (See Appendix, page 165.)

Darwin develops some of his most effective arguments in the chapters on the difficulties that might impede acceptance of his ideas. I think he is especially effective in handling the problem of the lack of intermediate forms, the rarity of the fossilization process in nature, and the universal sterility/fertility question. Yet, even here, difficulties remain. The lack of transitional forms, for example, although a problem for Darwin's theory, might be used to bolster the Creationist view. The Creationists could argue that transitional forms do not now exist because, in fact, they had never existed; the Deity presumably saw no purpose in their creation.

In discussing the lack of fossil remains of intermediate forms, Darwin can only appeal to the reader's ignorance and the imperfect state of the geological record. If the reader accepts Darwin's explanation of instincts and complex and useless organs, he must also accept several as yet unproven concepts: the existence of numerous intermediate gradations between organisms, the notion of variability in nature, and the concept of a struggle for existence in nature in which profitable variations are preserved. Here is another instance where phenomena such as the existence of instincts and complex and useless organs, although a problem for Darwin's theory, might be viewed as "evidence" in support of the Creationist view, in which every organism is presumably created with special endowments by an all-powerful Deity.

In his treatment of the problem of wide geographical diffusion of the same species with no apparent means of dispersal, Darwin is forced again to appeal to our state of ignorance of what might have happened; now he hypothesizes a glacial period in the distant past. The argument of wide diffusion is another instance where a difficulty for Darwin's theory might be viewed as support for the theory of a Creator who can produce similar species anywhere regardless of geography or barriers to migration. (See Appendix, page 166.)

Darwin's contrast of the explanatory powers of his theory with the Creationist, especially in the areas of geographical distribution, morphology, embryology, and rudimentary organs, represents, I think, the strongest line of arguments in the *Origin*. This is especially true of Darwin's treatment of these subjects in Chapters X through XIV of the *Origin*. These arguments match the idea on method and evidence Darwin articulated in the notebooks: the value of a theory is directly related to the number of diverse facts it can bring together under some general law or laws. At this point the logic of the Creationist position falters most, and the explanatory usefulness of Darwin's theory becomes most evident.

Yet even here, where Darwin's arguments are strongest, nagging questions remain. For example, a reader of the *Origin* might be justified in wondering what Creationist view Darwin is referring to. Perhaps this is a problem more for the present-day reader. Darwin's contemporaries may have known exactly what he meant, though I doubt it. Often the Creationist position seems merely a straw man—set up only to be knocked down.[122] The constraints on space in the *Origin*, which led Darwin to abandon his original intention of arguing on both sides of the mutability issue, add to this feeling. The result is that the Creationist position is never clearly defined in the *Origin*.

Although Darwin took great pains to contrast his theory with the Creationist view as an explanation of various natural phenomena, he only rarely contrasted his theory with other theories of evolution—Lamarck's, Erasmus Darwin's, the theory of the *Vestiges*, and others.[123] A reader might justifiably wonder if Lamarck's theory, for example, could explain most of the facts of nature as well as Darwin's. I do not think that Darwin's failure to deal in detail with other theories of evolution can be attributed to constraints on space in the *Origin*. There is no such consideration in the Long Manuscript, where space was more plentiful. Darwin probably eschewed such discussion in order to concentrate on what he viewed as his theory's chief competitor, the Creationist view. In the long run it was a wise strategic decision. A consideration of a whole range of theories would have overcomplicated an already complicated presentation.

Simply because Darwin's theory can explain certain natural phenomena better than the Creationist viewpoint does not mean that Darwin's theory is correct. It merely accords better with these particular phenomena. In such instances Darwin's theory is not necessarily the correct theory but only the least objectionable one.

I realize that Darwin need only present a less objectionable theory in order to have a better theory; I do not wish to imply a qualitative judgment on this argument or to suggest that others did not argue in much the same way. What I want to stress is that, given the relative paucity of evidence then available to Darwin, he was forced to rely heavily on the least objectionable form of argument. The term *least objectionable* is mine, but the strategy, as is evident throughout the *Origin*, was Darwin's. And the argument was not one about which Darwin felt very confident. A least objectionable theory was not what Darwin had hoped to communicate to the world.[124] (See Appendix, pages 167–69.)

An Interim Product

Given the limitations on space under which Darwin was working, the complex nature of many of the topics he was considering, our ignorance of so many of the phenomena he used to support his theory, the potentially fatal nature of the various objections that could be raised against his ideas, and the scant evidence Darwin was able to provide to support some of his most basic premises, I think a reader might justifiably wonder how Darwin could conclude anything about the origin of species. That Darwin was able to say something is a mark of his considerable achievement. Yet, the impression of a strong, but not completely comfortable, interim product lingers.

In this context, Darwin's reluctance to rush into publication or to publish a preliminary essay makes sense. He was not afraid of publication; he was unready for it. Instead of actually delaying publication, as many Darwin scholars have maintained, I think that Darwin was forced by circumstances to publish too soon. I suspect another year or two of work on the Long Manuscript would have produced a product more to Darwin's liking.

8

Darwin as Scientist-Manager

New Conclusions

In the introduction to this essay, I suggested that concentrating on Darwin's work on his species theory after he read Malthus would lead, in the course of our analysis, to several conclusions quite different from those now generally held regarding Darwin and his work. I think that the evidence discussed in the body of this essay supports the soundness of this prediction.

The year 1838 was not an end but an important beginning for Darwin. We have seen how rudimentary Darwin felt the initial formulation of his theory was; it was a "theory by which to work," and he "determined not for some time to write even the briefest sketch of it"; his desire to avoid "prejudice" was related not only to his fear of adverse public reaction to materialist ideas, but also to the weakness of his theory *qua* theory. He had produced, in effect, an uncertain theory on a highly controversial subject. Darwin's theory had really preceded his knowledge; in early letters to Fox and Jenyns he spoke of his species work quite tentatively. He worked on his theory, when he could, during the busy time following the *Beagle* voyage. His work included sorting out and finding a home for his innumerable specimens, arranging a volume on the zoological results of the voyage and securing a Treasury grant for the project, preparing for publication his *Journal of Researches* as well as several volumes on the geological results of the voyage, composing and presenting before learned societies a number of papers, mostly on geology, about the results of the voyage, and serving as one of the honorary, though working, secretaries of the Geological Society of London.

We have seen that Darwin's interest in natural history before the *Beagle* voyage—at Shrewsbury School, Dr. Butler's, Edinburgh, and Cambridge—had been casual. Natural history subjects during this time vied for attention with his beloved shooting; he was just beginning to study geology seriously when he accompanied Sedgwick on a geological tour of North Wales in the summer following graduation from Cambridge, and he was still very much a student of the subject during the *Beagle* voyage, though by then his enthusiasm for geology had increased dramatically. Henslow's recommendation of Darwin as naturalist aboard the *Beagle* was certainly a qualified one, and during the voyage Darwin was primarily a collector for the "lions" back home, as he himself admitted in a letter to his Cambridge friend Herbert. His field researches on the voyage were directed in large part by Henslow, as Darwin later publicly acknowledged; upon his return from the voyage he depended on the help of others to sort and classify his specimens. He sought "experts" for his volume on the zoological results of the voyage and needed Henslow's help in putting together his *Journal of Researches.* Given the rudimentary state of his knowledge at the time, Darwin felt that he ought to have prepared detailed zoological and geological volumes before attempting to put together the more general *Journal;* he realized he was at the beginning of his career in natural history and indicated so in the letter of October 14, 1837 to Henslow concerning the Geological Society post. His species notebooks reveal Darwin as a person with some knowledge seeking more, as a neophyte naturalist at the beginning of his investigations into the species question.

With respect to the part-time nature of Darwin's species work, we have seen that his primary interest during the *Beagle* voyage was geology, not zoology or botany or species; following his return from the *Beagle* voyage he first attempted a career in geology, which he actively pursued for over a decade, during which time he laid aside his species work, as his *Diary* reveals. He left geology only when his career in that field seemed at a standstill; he did not then return full-time to his species work, but rather turned to another area of investigation, taxonomic zoology, an area of only indirect importance to his theory.

Darwin depended on others and was able to benefit from the

help that others offered. We have seen how Hooker aided Darwin by answering questions, making observations, performing experiments, reviewing manuscripts, criticizing ideas, and performing a host of other supportive services; how Darwin depended on a whole group of people for similar support; how putting the *Origin* together became as much a matter of organizing and managing people as formulating new ideas. Darwin was able to use his organizational and management skills, his acute sense of the value of time, his personal charm and wit, his ability to persuade, and his perseverance to get others to work for him. He gradually evolved a world-wide correspondence network, with questions going out and specimens, observations, and data coming in, to gather the information he needed.

Regarding Darwin's doubt and uncertainty about his theory, we have seen the difficulty he had clearly explaining his ideas to others; the difficulty he had at times grasping his own ideas; the often overwhelming nature of his subject; his inability to provide scholarly references; his fear of negative public reaction and resulting loss of stature in his profession; and his concern about his failure to convert many of his contemporaries to his ideas. On the eve of the publication of the *Origin*, only Wallace, the co-discoverer of the theory, and Hooker, a recent convert after many years of skepticism, could be counted as standing firmly on Darwin's side.

Finally, with regard to the quality of the theory itself, we have seen Darwin's doubts as expressed in his letters; his concern with substantive weak spots in the pre-*Origin* material and with strategic questions of how best to communicate and argue his case. He realized that several important questions of a "common sense" nature could be raised against his theory, and that in the *Origin* his argument rested, in many important instances, on new and often unsubstantiated hypotheses, on inexact analogies and metaphors, on the repudiation of competing explanations, and on a frequent plea to ignorance and complexity, rather than on substantial, clearly incontrovertible evidence in its own support.

In sum, if Darwin did not have a finished theory in 1838, he had only an interim product by 1859. If 1838, as I have argued, did not mark an end, but an important beginning for Darwin in the development of his theory, then 1859 did not mark an end either, but rather an important mid-point. Succeeding editions of the *Origin* and the publication of several other species-related works by Darwin—*Variation of Animals and Plants Under Domestication* (1868), *The Descent of Man* (1871), to cite two of the more important—were needed to present Darwin's theory in a more finished form before the public for review.[1]

Darwin's Personality

Understanding Darwin's accomplishments during the period 1838–59 requires some insight into Darwin's personality. Two characteristics stand out: Darwin's ambition and his acute sense of reality.

I have already discussed Darwin's ambition in some detail in Chapter III. Ambition implies nothing pejorative about his character. On the contrary, I think that to put together a work of the *Origin*'s magnitude required strong ambition. Nevertheless I do not wish to hide his ambition; it needs no apology.

Darwin undoubtedly wanted to make a major contribution to the corpus of scientific knowledge. I do not think that it mattered to Darwin, especially as a young man, where he achieved recognition: in geolgy, zoology, or evolutionary theory. The important thing was that he achieve some noteworthy place, some "fair place among scientific men," as he put it.[2] The problem was that his standards for success were very high.

Although his contributions to geology were considerable, by the mid-1840s Darwin felt his career in this area was at a standstill. The sparse sales of his works irritated him, and his ill health interfered with continued geological field research. Other contributing factors were the powerful presence of Lyell in geology at the time and the fact that by 1846 Darwin had overseen preparation of nearly all his South American geological materials.

Darwin found perhaps even less satisfaction in his work on barnacles, which he took up next. Although initially intended as a quick look at one special problem related to the Cirripedes, Darwin's work in this area dragged on for eight years, during which time his enthusiasm waned. In a letter on March 28, 1849 to Hooker, Darwin talked about how happy he would be to leave hydropathy treatments at Malvern and return to Down to resume work on his "beloved Barnacles."[3] A few years later, in a letter dated October 24, 1852, Darwin told Fox about his work on "the second volume of the Cirripedia, of which creations I am wonderfully tired. I hate a Barnacle as no man ever did before, not even a sailor in a slow-sailing ship."[4] In March 1855, Darwin wrote, again to Fox, to say how glad he was to be "at last quite done with the everlasting barnacles."[5] Years later, in his *Autobiography*, Darwin expressed doubt over whether his work on Cirripedes "was worth the consumption of so much time."[6] One problem Darwin must have felt about his work was its rather narrow focus. It had none of the excitement of the "grand ideas" of science that Darwin associated with geology and his work on species.[7]

In November 1853, Darwin received the Royal Medal of the Royal Society of London. This was an important award, though not as prestigious as the Royal Society's Copley Medal.[8] According to the statement read by the Earl of Rosse, then president of the Royal Society, the award was given to Darwin for his work on coral reefs and his Cirripedes studies. There was also a short reference to Darwin's collection of facts and careful observations in support of Lyell's theories.[9] Although Darwin was quite pleased with the award, it must have been a somewhat bittersweet and ironic moment for him; he had years before abandoned any active career in geology, and by that time he was weary of barnacles.[10]

If Darwin could not find satisfaction in a career in geology or in his tedious, detailed barnacles study, what was left? In May 1854, Darwin took up species full-time. By June 1858, he had made considerable progress with his work, but all his hopes for recognition seemed dashed when, on June 18, 1858, he received Wallace's letter.

A number of Darwin scholars have described what happened following Darwin's receipt of Wallace's letter: at the urging of Lyell and Hooker, who both thought that Darwin should share equally with Wallace in the discovery of the theory, a reluctant Darwin agreed to a joint Darwin-Wallace paper to be read before the Linnean Society of London. The paper consisted of an extract of Darwin's essay of 1844 and a letter he wrote to Asa Gray in 1857 describing his theory, together with Wallace's paper. The joint paper was read on July 1, 1858, and was published in the Society's Journal on August 20, 1858.[11] Darwin's priority was thus assured, and fifteen months later the *Origin* appeared.[12] One scholar has described this episode as involving "that mutual nobility of behavior so justly celebrated in the annals of science."[13]

Although nobility was not completely absent from this episode, I have a slightly different interpretation of what happened. In a letter to Lyell sent June 18, 1858, the same day Darwin received Wallace's letter, Darwin made two things very clear to his friend: one was that Wallace did not raise the question of publication of his ideas—"he does not say he wishes me to publish"—although Darwin added gallantly that he felt he ought to arrange for publication anyway. The other was Darwin's grave disappointment at the turn of events—"so all my originality, whatever it may amount to, will be smashed"—though he added that the value of his work will not be diminished much, for the value was in the application of the theory. It is obvious from the letter, though, that Darwin was distraught over what had occurred.[14]

Only in response to this letter did Lyell suggest that a joint publication of the theory be arranged. Hooker was not yet involved. On June 25, 1858, Darwin responded to Lyell's suggestion, again making clear that Wallace had said nothing about publication (as proof he enclosed Wallace's letter). Darwin also made Lyell understand that there was nothing in Wallace's sketch that he had not written about more fully years before.[15] Although Darwin says in the letter that his first impression was that it would be paltry for him to try to publish anything now, he apparently had given second thoughts to Lyell's suggestion of joint publication. Darwin concludes the letter by suggesting that Lyell might want to send Darwin's letter, together with his own recommendation for joint publication, to Hooker for his opinion in the matter.[16] The next day (June 26, 1858) in a follow-up letter to Lyell, Darwin candidly admitted that " . . . it seems hard on me that I should be thus compelled to lose my priority of many years' standing . . . ," though he was still concerned about the justice of the case.[17] Hooker, with Lyell's suggestion and Darwin's letter of June 25 in hand, agreed with Lyell's idea, and the joint paper was prepared for presentation to the Linnean Society.

I do not wish to delve too deeply into the ethical questions associated with this matter. I wish to point out only that it seems that Darwin, quite humanly, was determined to make certain that he received his fair share of the credit for discovery of the natural selection theory, and the idea for joint publication was not something that had to be forced upon him. I think that he hoped Lyell or Hooker might intervene in some reasonable manner in his behalf. Darwin was determined to make Lyell (and through Lyell, Hooker) aware that Wallace did not request publication of his ideas, that he was upset over his pending loss of priority, and that he would be willing to accept, albeit after some moral anguish, Lyell's suggestion about joint publication.[18]

Several Wallace scholars have strongly implied that Darwin behaved in a less than admirable way in this episode.[19] In Darwin's defense, one could argue that this period in his life was a particularly difficult one for him: he was more ill than usual, and only a few days after he received Wallace's letter, an infant son had died. To add to Darwin's worries, his daughter was then seriously ill with diphtheria. Darwin was depressed at the time, and it is obvious that he found the whole episode rather unpleasant. But there can be no doubt that his instincts to protect his species theory had been aroused.

A further indication of Darwin's determination to receive his fair share of priority is that soon after the Linnean presentation Darwin stopped work on his Long Manuscript and began to prepare a short abstract of his views, for immediate publication. Given Darwin's penchant for thoroughness and detail, this work soon became a rather long abstract, entitled *The Origin of Species.*

Why did Darwin feel compelled to rush an abstract into publication? The Linnean joint paper certainly assured him co-priority with Wallace for discovery of the natural selection theory. Did Darwin feel that Wallace was about to publish substantial work supplementing his June 18, 1858 sketch?[20] I do not know. I assume that Darwin wanted to bolster his priority for the theory and was willing to sacrifice some degree of quality for quickness of publication.

Robert Stauffer's work on Darwin's Long Manuscript is illuminating in this regard. Darwin began writing the Long Manuscript on May 14, 1856. When Wallace's letter interrupted Darwin's work on June 18, 1858, Stauffer tells us, Darwin had covered roughly two-thirds of the topics later covered in the *Origin.* Of the fourteen chapters of the *Origin,* nine were treated at length in the Long Manuscript. Stauffer estimates the completed long work would have been 375,000 words, with 225,000 words already finished by June 1858. Thus, at the time Darwin stopped working, he was approximately 150,000 words short of completing the larger work. Also, Stauffer tells us, the Long Manuscript, at that point, had progressed far beyond first draft stage. Many chapters had been revised, rewritten, reorganized, expanded, and supplemented. I agree, as I mentioned earlier, with Stauffer's contention that, compared with the *Origin*, the Long Manuscript has more examples illustrating Darwin's theory as well as being a fully referenced work. It is simply a much stronger, scholarly product.[21]

Ironically, it took Darwin fifteen months to finish the *Origin*. The *Origin* ran to 155,000 words. Thus, in terms of length at least, Darwin could just as easily have completed the long, fully referenced version of his theory. Nor is there much evidence to support the idea that a work longer than the *Origin* would have been any less salable or had any less impact.[22] Stauffer points out that " . . . the scale [of the long version] does not seem inordinate considering the standards of the days of double-decker and triple-decker novels."[23]

So much for Darwin's ambition. Tempering it and leading it into particularly useful and creative channels was the second characteristic I mentioned—his acute sense of reality. Many people want to achieve recognition; few understand how. Darwin was one of these few. He may have had fantasies about what he could accomplish, but they do not seem to have influenced his actions. Darwin always seemed to sense where he stood in a given project, whether it was geology, Cirripedes, or species. He knew what the problems or questions were, what he needed to learn, who could help him, and, perhaps most importantly, how to get help.

When Darwin began sending back specimens from the voyage and receiving praise of his work, he knew that he was just a collector for the others. Upon returning from the voyage, he knew he would have to get his specimens to the right people if they were to be used wisely; he knew he needed Henslow's help to get his *Journal* out; he knew he needed experts for his zoology volume.

In 1838, Darwin realized his theory was still very preliminary; he knew that he did not have adequate evidence to support it. Undoubtedly, as his letter to Emma on July 5, 1844 concerning disposition of his theory reveals, he felt he had hit upon an idea of first importance in natural history, yet he knew he was in no position to do much about it. He would collect more facts for his species work and see how his career in geology progressed.

Throughout the late 1830s and 1840s Darwin's illness interfered more and more with his available working time. Darwin knew that productivity would depend on periods of relaxation interspersed with periods of work. He also knew under the circumstances that it would be impossible to do his species work by himself; he would have to enlist the help of others.

When he finally did turn to species work full-time in 1854, he had a good sense of what was required to get his theory into presentable shape. He started to piece together, in the most methodical and deliberate way, a substantial work on species. He resisted Lyell's urgings to publish a preliminary essay.[24] Darwin understood that a preliminary sketch would probably invite more abuse than praise; he knew he had not yet grasped the full implications of his own theory.

Darwin had the uncanny ability to stand back, objectively analyze his situation, and then either move with great purpose and direction into action, to perform an experiment, get a study, request a specimen, have an observation made, send a manuscript off for critical analysis, or do nothing, simply defer action.

He had an innate sense of, and aptitude for, planning and managing an enterprise—in my opinion, the quintessential element in all successful management efforts. He seemed to know instinctively what to do under a variety of circumstances and to be able to make the most of a given situation. He had what we today would call good "street sense." Problems for Darwin, rather than barriers to progress, were the bases of new courses of action. Ignorance led, not to frustration, but to bibliographies of books and articles to be read. New and exciting ideas stimulated hypotheses to be tested and retested. Imagination, he made certain, was countered with reality, impetuosity with caution. Not by chance the bold-thinking Darwin chose the skeptical, practical Hooker as a partner in his species work.

Darwin as Scientist-Manager

Darwin was a scientist-manager par excellence—one who could not only develop new ideas but organize and communicate those ideas and convince others of their validity. Darwin had a sense of his audience and a feeling for the arena of competing scientific theories; he had the capacity to plan a major research project, devise a long-term agenda for that project, and create, organize, and maintain an informal world-wide research organization to support that project. He also had the ability to manage and organize his own life and work, devising, as we have seen, a rigid schedule of work and relaxation in order to ensure that every ounce of available energy would be directed toward completion of his task.

Finally, Darwin was a person whose organizational and management skills extended to preparation of the product itself, where he showed great skill at marshalling evidence, structuring arguments, using whatever facts, ideas, notions, hypotheses, and analogies he could find to bolster his case. Dawin had an unusual talent which combined the creativity of a bold, imaginative thinker with the strategic and organizational capabilities of a highly competent manager and entrepreneur. It is these capabilities, which until now have received little notice, that I see as crucial to the development of his work.[25]

An Imperfect Process

Some additional thoughts on Darwin and his theory: one fascinating aspect of the overall process by which Darwin developed his ideas was that almost nothing about it was simple, neat, or easy. It was, in fact, a rather fragile, profane, and imperfect process. For twenty years Darwin worked intermittently on his theory. Although there was one single moment of great inspiration—after Darwin read Malthus—there were many other moments of lesser inspiration—discovering the principle of divergence of character, understanding the means of organic transport and dispersion, and untangling the problem of the imperfection of the geological record. There were probably numerous other moments of absolutely no inspiration.

Many extraneous things usually not associated with the scientific creative process seemed to affect the process—Darwin's salesmanship, his management and organizational skills, his strategic capabilities, his charm. These seemed to be, at times, as important as the development and refinement of his ideas, especially as Darwin was so dependent on the help of others to keep the process going.

There were many other things that could, at any moment, slow down the process or even stop it altogether. It always seemed close to breaking down. What would have happened if Darwin's career

in geology had matched his expectation? Would that have pre-
cluded his species work? Or what if his work on Cirripedes had
opened new theoretical vistas in the taxonomy of organisms?
Would eight years have been just the beginning? Or what if Lyell
and Hooker had not interceded in his behalf in the Wallace priority affair? Would Darwin have been completely demoralized?
Would he still have had the enthusiasm to proceed with publication?

The process was also very imperfect and untidy in human terms.
Darwin had to enlist the support of others to keep going. Although
he felt guilty about asking for help, his isolation at Down gave
him no choice. Hooker, Fox, and Gray seemed not to mind Darwin's requests; they were fairly willing helpers. But what of the
others? Did politeness prevent them from complaining? Also, Darwin obviously found the whole Wallace affair very unpleasant.
He had no intention of publishing at the time he received Wallace's letter.[26] Although Darwin had no role in the final arrange-
ments and hence cannot be held responsible for them, the final
product of the affair—the joint Linnean paper—was, in my opinion, heavily weighted on Darwin's side to give him the lion's share
of credit for the theory. Darwin was certainly surprised. He had
expected his material to be merely an appendage to Wallace's
sketch, but just about the opposite happened: Wallace's work followed the extract of Darwin's 1844 essay and his 1857 letter to
Gray in the joint paper.[27]

Finally, there was the unsatisfactory interim product of this
overall process—Darwin's *Origin of Species.* Quickly put together
under tremendous pressure, it must have seemed to Darwin more
rhetorical than evidential in nature; it must often have appeared
to derive its greatest strength from what other theories could not
explain, rather than from what his could.

But perhaps what I have described here is how the creative process usually works; perfectly clear and well-arranged ideas come
simply and neatly out of the minds of heroes and angels only in
distant retrospect.

Previous Theories About Darwin

How do my ideas about Darwin fit in with previous scholarship in the area? This is a difficult question to answer, primarily
because of the enormous amount of literature produced on Dar-

win in the last thirty years, and especially since the centennial of the publication of the *Origin* in 1959. Yet it is a particularly important question in a field that is growing so rapidly.

John C. Greene, in his review of secondary literature, "Reflections on the Progress of Darwin Studies,"[28] attempts to categorize the enormous corpus of Darwin scholarship by dividing it into three basic groups: studies by intellectual historians, studies by scientists interested in history, and studies by scholars specifically trained in the new academic disciplines of the history and philosophy of science.[29]

Intellectual historians originally came to the study of Darwin following in the footsteps of such scholars as Arthur O. Lovejoy, A. N. Whitehead, and R. G. Collingwood. Lovejoy, in particular, helped focus attention on Darwin in his articles on eighteenth century evolutionists and in his classic work, *The Great Chain of Being*.[30] More recently, in the last ten to fifteen years, many intellectual historians interested in Darwin and related subjects have migrated to the new fields of the history and philosophy of science.

Greene describes the intellectual historian's approach to Darwin as primarily concerned with the relationship between Darwin's ideas and those of his predecessors and contemporaries and with the influence of Darwin's writings on Western thought. The intellectual historian has tended to view Darwin primarily as a thinker within the social and intellectual context of his times. As a result, he has tended to disregard the basis of scientific evidence and argument in Darwin's theories. Also, the intellectual historian's focus on sources has not gone much beyond the *Origin* and the *Descent of Man*.[31]

Some of those Greene places within this group include Jacques Barzun, Bert James Loewenberg, Charles Gillispie, A. Hunter Dupree, Gertrude Himmelfarb, Donald Fleming, Robert M. Young, and Greene himself.[32]

Scientist-historians came to a study of Darwin from specific science backgrounds. To Greene, much of the progress that has been made in analyzing and evaluating the scientific work of Darwin, his predecessors, and his contemporaries can be attributed to the scientist-historian's work. Unfortunately, many scientist-historians have approached Darwin in a highly polemical way, with considerable ideological baggage, "carrying the scientific

quarrels and terminology of . . . [their] own times back into the
past and seeking to make Darwin take sides in controversies which
were yet unborn.[33] The zoologist Michael Ghiselin is cited as a
scientist-historian particularly guilty on this score.[34]

There have been, according to Greene, a few scientist-historians
who have raised themselves above such quarreling and have
"shown themselves capable of solid scientific analysis with a due
appreciation of the complexities of intellectual history"—people
such as Ernst Mayr (on Lamarck), Martin Rudwick (on the history of paleontology), and Howard Gruber (on Darwin).[35] In addition to Ghiselin, Mayr, Rudwick, and Gruber, Greene includes
within this group such scholars as S. A. Barnett, P. R. Bell, C. D.
Darlington, Gavin de Beer, Loren Eiseley, and George Gaylord
Simpson.[36]

Scholars coming out of the recently expanded disciplines of the
history and philosophy of science comprise the third and largest
group of those professionally interested in Darwin and related subjects. These scholars have covered the whole gamut of topics related to Darwin, his precursors, and his contemporaries.

Greene identifies the French and French-Canadian School of
Georges Canguilhem, Camille Limoges, Yvette Conry, and Jean-
Claude Cadieux.[37] In the British/Cambridge group, Greene iden-
tifies such scholars as Robert Young, Martin Rudwick (at Cam-
bridge originally, now at Edinburgh), Jonathan Hodge, William
Bynum, and Michael Bartholomew.[38]

In the United States the list is vast. To mention but a few by
area of interest: Darwin's precursors—Richard Burckhardt (on
Lamarck), Paul Farber (Buffon); Darwin's contemporaries—H.
Lewis McKinney (Wallace), Leonard Wilson (Lyell); Darwin and
his work—Peter Vorzimmer (Darwin's changing theory), Robert
Stauffer (the Long Manuscript); the intellectual setting—Susan
F. Cannon (Lyell, Victorian science); systematic accounts of var-
ious fields of natural history in the nineteenth century—William
Coleman (biology), Jane Oppenheimer (embryology), George W.
Stocking, Jr. (the natural history of man), and so on.[39]

Not all individuals fall easily into only one group; for example, Greene describes Robert Young as both an intellectual historian and a historian of science; I would include Susan F. Cannon
as an intellectual historian (which Greene does not) as well as a
historian of science (which Greene does), but Greene's categori-

zation is useful as the heuristic organizing device he obviously intended it to be.

Whatever the origins of their interests or their particular professional training, there has been a wide variety of opinion among scholars about Darwin and his work. Jacques Barzun, one with a negative opinion, sees Darwin as a "modest, industrious, and wholehearted man,"[40] but "preeminately an observer and recorder of facts."[41] Facts impinged on Darwin's mind far more deeply and significantly than abstractions.[42] Darwin "came to ideas and he came by ideas very slowly."[43] Although Darwin considered himself an innovator and an iconoclast, Barzun feels he worked most of the time with other people's assumptions and ideas.[44] Though Darwin was "a great assembler of facts," to Barzun he was nonetheless "a poor joiner of ideas."[45] He had a confused habit of mind which repeatedly led him into tautology,[46] was often hazy, and expressed himself "darkly."[47] Even when he begged a question, which he did often, he did not do it very well, "because the begging generally covers pages of circumlocutory matter."[48] Darwin was not only obscure; he also hedged so much and was so self-contradictory that he enabled people to quote the *Origin* as one might quote the Bible—for one's own purposes.[49]

To Barzun, Darwin simply did not have the power of sustained intellectual thought; "Darwin was not a thinker and he did not originate the ideas that he used. He vacillated, added, retracted, and confused his own traces. As soon as he crossed the dividing line between the realm of events and the realm of theory he became 'metaphysical' in the bad sense."[50] In short, "Darwin . . . does not belong with the great thinkers of mankind."[51]

Another detractor, C. D. Darlington, sees Darwin as having skill and persistence as an inquirer, observer, and recorder and having a certain independence of thought, but also as an intellectually opportunistic, cowardly sort of person, who took certain ideas from his predecessors, the "medical evolutionists" James C. Prichard and William Lawrence, did not acknowledge his debt, and then presented a theory "a little loose in its . . . arguments" in order to enable himself to hide if attacked.[52] Darlington argues that Darwin quotes Lawrence five times on matters of detail, but never on the main issue of evolution. He "does not refer to Lamarck's principle even to repudiate it."[53] Darwin's "self-centeredness," Darlington continues, made him unaware of what his predecessors had written.[54] Darwin also confounded theoretical issues, concealing the confusion of his ideas "from his indiscriminating readers by loose writing or . . . by making a mystery wherever it was wanted."[55] In short, according to Darlington, Darwin's vices were helpful in his success, one of which was "a flexible strategy which is not to be reconciled with even average intellectual integrity: by contrast with Wallace, Lyell, Hooker, Chambers or even Spencer, Darwin was slippery."[56]

Somewhat less negatively, Gertrude Himmelfarb sees Darwin as a single-minded, hard working scientist, sensitive to the needs of his friends (he helped get a government pension for Wallace; he helped support a subscription to enable Huxley to manage his affairs after a particularly serious illness);[57] but she also sees him as "less ambitious, less imaginative, and less learned than many of his colleagues," one whose "essential method was neither observing nor the more prosaic mode of scientific reasoning, but a peculiarly imaginative, inventive mode of argument."[58] To Himmelfarb, Darwin created what she calls a "logic of possibility," but "unlike conventional logic where the compound of possibilities results not in a greater possibility, or probability, but in a lesser one, the logic of the *Origin* was one in which possibilities were assumed to add up to probability." Himmelfarb writes:

[The] . . . technique for the conversion of possibilities into probabilities and liabilities into assets was the more effective the longer the process went on . . . the reader was put under a constantly mounting obligation; if he accepted one explanation, he was committed to accept the next. Having first agreed to the theory in cases where only some of the transitional stages were missing, the reader was expected to acquiesce in those cases where most of the stages were missing, and finally in those where there was no evidence of stages at all. Thus, by the time the problem of the eye was under consideration, Darwin was insisting that anyone who had come with him so far could not rightly hesitate to go further. . . . As possibilities were promoted into probabilities, and probabilities into certainties, so ignorance itself was raised to a position only once removed from certain knowledge.[59]

Himmelfarb feels that Darwin's achievement was enormous because it was revolutionary in effect, but she seriously questions the quality of that achievement and the capabilities of the person who produced it.[60]

More positively, Loren Eiseley sees Darwin as a brilliant observer of nature who synthesized a whole range of scientific ideas gathered from Lamarck, Malthus, Lyell, Edward Blyth, and others to form a comprehensive theory of nature—really a quite remarkable achievement.[61] But Eiseley also sees Darwin as being indifferent "to the history of the ideas with which he worked." Eiseley believes "great acts of scientific synthesis are not performed in a vacuum." According to Eiseley, Darwin had a tendency to emphasize the antagonistic elements prevalent in nature to the detriment of the cooperative ones; that is, he failed to see that organisms cooperate as much as they struggle and that there is a certain internal bodily harmony within organisms.[62] Eiseley also sees Darwin, in later editions of the *Origin*, backing off from certain earlier positions in response to various criticisms, to the detriment of his overall theory.[63]

Another supporter, Charles C. Gillispie, sees Darwin as the developer of a "new natural philosophy, as new in its domain as Galileo's in physics."[64] Darwin "treated scientifically the historical evidence for evolution, which had been marshalled often enough before him, but more as a travesty than an extension of science."[65] Instead of explaining variation, "he begins with it as a fundamental fact of nature,"[66] thus allowing objectivity to enter the life sciences and paving the way for the eventual mathematization of these fields.[67] In this effort, Darwin was without predecessors, and his contribution essentially novel. Also, Gillispie, unlike Darlington, sees Darwin as a person of considerable moral courage, unafraid to pursue potentially controversial ideas or face "the unknown four-square, taking only science for his guide."[68]

Michael Ghiselin recognizes Darwin as "a scientist of the first rank,"[69] whose methodical working habits, independence of thought, strength of character, and self-esteem played no small role in his success. Yet Ghiselin feels that these factors cannot by themselves explain Darwin's achievement. He maintains that Darwin "applied, vigorously and consistently, the modern, hypothetico-deductive scientific method" and that without understanding this, the full extent of his contribution cannot be fully appreciated.[70] Darwin's success resulted from the nature of this hypothetico-deductive method, which he applied consistently throughout his mature works. In this sense Darwin's achievement was not so much the collection of facts as the development of a very sophisticated theory.[71]

Confronting Darwin's detractors, Ghiselin suggests that far from plagiarizing his ideas (as Ghiselin claims Eiseley had charged with regard to Edward Blyth), Darwin developed novel and brilliantly reasoned explanations for a variety of natural phenomena, from the origin of coral reefs to the origin of species.[72] Although unconscious of the method he was using, Darwin was nonetheless essentially modern in his approach.[73]

Julian Huxley, George Gaylord Simpson, and Gavin de Beer, among others, have also strongly stated their support of Darwin.[74]

There is another group of scholars who, while not refraining from taking strong stands on various issues, have avoided polemical positions regarding Darwin and have concentrated instead on trying to understand the nature of the man and his work.

Susan F. Cannon is one of this group. Cannon sees Darwin's work not embodied, like Charles Lyell's, in one theoretical framework or scheme, but composed of concepts from various sources: English ways of being scientific (the English tradition of natural theology); a Humboldtian vision of nature (topographical principles at work in time and space); the theoretical methods of Charles Lyell, especially the use of systems of hypotheses and deductions and Lyell's idea of small changes adding up to large effects over time; eighteenth-century ideas of evolution, transmutation, utilitarianism, and so on.[75] She sees a definite uniqueness to Darwin's ideas; Darwin's theory was not, given the scientific climate of opinion of the 1850s, the obvious next step.[76] In fact, despite what Darwin gained from Lyell, Cannon sees Lyell as Darwin's chief antagonist; the Uniformitarian theory, after all, did not allow for progressive development of any sort, and Lyell devoted a whole section of the *Principles* to a refutation of Lamarck's evolutionary views.[77] Darwin's basic problem was one of constructing a reductionist (materialist) system which would explain the origin of man's "higher" faculties without offending the sensibilities of those who believed in natural theology, that is, of those who, unlike Uniformitarians, were willing to admit some sort of progressive development of organisms in nature.[78] Unlike most scholars, Cannon does not believe Darwin delayed publication of his ideas. "His conduct was professionally valid. He had no proper scientific case in 1839, only a set of speculations; he needed twenty years or so to develop one. Even then it came out, under pressure, in bits and pieces. . . . "[79] Finally, Cannon regards the first edition of the *Origin* as somewhat flawed, not very logically or rigorously constructed. In it Darwin was trying to convince, not to prove. "He was trying to get across his own complex vision of how the natural world operates; and he seized upon any metaphors, any analogies, any line of argument that he had on hand."[80] Seeing Darwin as not particularly clever or bright, Cannon describes him as having the tendency to jump to conclusions "well beyond the evidence available" and maintain "his faith in his position regardless of the valid arguments that could be brought against it."[81] She feels, however, this intuition combined with Darwin's stubbornness carried the day for his theory.

Many other Darwin scholars fall within this neutral, or at least, non-polemical, category. To mention a few: Howard Gruber, who describes Darwin's groping to develop his theory, delineates the period between Darwin's reading of Malthus and the publication of the *Origin* (1838–59) as one of puzzling delay. He describes the fear with which Darwin viewed the materialistic implications of his theory and suggests that Lyell taught Darwin to read well potential opposition and trim his argumentative sails accordingly.[82] Peter Vorzimmer attempts to show that Darwin had become, "under critical attack, increasingly frustrated by his inability to prove to the satisfaction of fellow scientists that the selection process was the sole or, in some cases, the principal agent of evolutionary development" and that Darwin, as a result, had to modify his theory piece by piece in successive editions of the *Origin* until it became substantially a different theory.[83] Camille Limoges discounts Darwin's search for an analogue to artificial selection, emphasizing instead the importance to Darwin of the concept of relative adaptation in nature.[84]

There are, of course, many other views of Darwin and his work. I have sought here to give merely a flavor of the views of some of the more prominent scholars.

In terms of Greene's three categories, I fall within the third group, those specifically trained in the history and philosophy of science. I would like to think that my views on the broad spectrum of opinion on Darwin fit within the neutral, or non-polemical, category, though my conception of Darwin has been influenced by views all across the spectrum. I agree, for example, with several of Barzun's ideas about Darwin. There were times when Darwin did not express himself very clearly, and Darwin, as we have seen, was well aware of this problem. As a result, there are some sections of the *Origin* that are rather difficult to understand.[85] I think Barzun's description of Darwin as "a poor joiner of ideas," however, misses an important aspect of Darwin's achievement: his ability to put together in reasonably effective fashion an incredibly complicated argument based on a relative paucity of evidence. A person who was only a great assembler or collector of facts could not do that.

I agree, in a sense, with one of Darlington's criticisms of Darwin —that his theory was "a little loose in its arguments," if Darlington means that Darwin did not always have the strongest evidence to present in support of his ideas and that at times he was forced to use weak arguments.[86] But I do not think that he consciously, as a matter of strategy, used weak or obscure arguments in order to hide behind them if attacked. I think, on the contrary, Darwin was trying to present the strongest case possible, and he used his strategic capabilities, which, indeed, were highly developed, to make his arguments as effective as possible, as is apparent from his species writings and notes before the *Origin*.

I believe that there is much to recommend Himmelfarb's attack on the quality of Darwin's scientific reasoning, in which possibilities were promoted into probabilities and probabilities into certainties; there is no doubt in my mind that in certain parts of the *Origin*, Darwin's argumentative reasoning leaves much to be desired. But I also believe that Himmelfarb has focused on these soft spots to the detriment of all else. There are many strong arguments in the *Origin*, as well as much very effective reasoning. I understand Himmelfarb's reaction. I remember my own shock when first coming upon some of Darwin's less convincing arguments in the *Origin*. It was difficult to reconcile them with the world-wide impact that his ideas have had. The soft spots seemed the more shocking because they were so unexpected.

Eiseley's point, which he shares with Barzun, Darlington, Himmelfarb, and many others, that Darwin fails to give due credit to his predecessors, has been overblown. I do not think it was Darwin's responsibility to give credit to his predecessors, although it would have been nice if he did. He had enough to do merely trying to put the *Origin* together, let alone trying to figure out where he got his ideas. When Darwin did have more time, in later editions of the *Origin*, he did provide a historical sketch. Perhaps that sketch was not very good, but the most one can accuse him of on this score is not being a good historian of science. I disagree with Eiseley's other point that Darwin tended to emphasize the antagonistic elements in nature to the detriment of the cooperative ones. I think Darwin tried to soften his vision of a struggle for existence by emphasizing the metaphorical nature of such struggle and the "dependency" in nature of one being upon another, though to do so successfully was difficult, given his stance in the *Origin* that previous naturalists had not seen the true intensity of a struggle for existence in nature.

Although I would not go as far as Gillispie does in divorcing Darwin's achievement from the work of his predecessors—Darwin simply owed too much to Lyell and others—I do agree with Gillispie's assessment that Darwin helped introduce objectivity into the life sciences, though he never really gave up the idea of attempting to understand the origin of variation—witness his theory of Pangenesis.[87] He was tactically smart enough, however, as we have seen, to realize that to hark back to the beginning of things would only weaken his argument.

I think I probably agree with Greene's overall assessment of Ghiselin's book: "a brilliant but profoundly unhistorical work."[88] Ghiselin has succeeded in showing the existence of Darwin's hypothetico-deductive method (I still question how often Darwin used it), though I think that Ghiselin seriously overrates this particular method as a key to understanding Darwin's success; as a result, Ghiselin also overrates Darwin's prowess as a theoretical thinker. I would maintain that Ghiselin commits Himmelfarb's mistake, but in the opposite direction: that is, whereas Himmelfarb sees some instances of unrigorous thinking in Darwin's work and quickly dismisses Darwin as a less-than-adequate thinker, Ghiselin sees some instances of quite modern thinking in Darwin's work and proclaims him a brilliant theoretician, many years ahead of his time. I do not see the basis for either characterization in Darwin's writings.

I have been impressed by Susan F. Cannon's rethinking of Darwin's period of long "delay," her description of the rather imperfect nature of the *Origin*'s argument in which Darwin was trying to convince, not to prove, and her ideas concerning Darwin's tendency to jump to conclusions "well beyond the evidence available," which I translated into the idea of Darwin's initial theory "preceding" his knowledge. Cannon's view of Darwin's ideas being composed of a number of concepts derived from various sources also impressed me, although I do not always agree with her on what those sources might have been.

Gruber's idea of Darwin's groping toward his theory is an intriguing one, and one which seems more plausible to me than Ghiselin's hypothetico-deductive theory view. Vorzimmer's idea of Darwin's retreat matches my view of Darwin as a scientist-manager with a grasp of reality, willing to modify his ideas in the face of mounting criticism, yet I do not think that Darwin retreated as much as Vorzimmer (like Eiseley before him) claims. Finally, I remain skeptical of the basis of Limoges' contention that Darwin did not rely on the domestic breeding analogy, unless Darwin's numerous references to it are to be ignored.

Appendix
A Summary of Darwin's Theory and the Evidence For and Against It

Major Elements of Darwin's Theory

Element	Evidence Presented in Support
1. Variation	Certainly exists in a domestic setting, at least to those practiced in domestic breeding; in nature it exists also; numerous variations occur, but hardly ever noticed by naturalists; Darwin indirectly answers question of existence of variation in nature by discussion of differences between species and varieties, showing that variations are constantly being produced in nature; we are ignorant about the laws of variation.
2. Struggle for Existence	Some naturalists (De Candolle, Lyell), have shown, but have not shown in true intensity; Darwin shows how a struggle for existence might happen (using Malthus as a point of reference) because of an imbalance between population growth and food supply; Darwin provides a few examples of struggle based on various experiments performed at Down.
3. Inheritance of Acquired Characteristics	Uses domestic breeding analogy; laws of inheritance unknown.
4. Divergence of Character	Uses domestic breeding analogy; provides a hypothetical example of how it might occur in nature.
5. Extinction	Assumed; indicates *Beagle* voyage evidence (remains of mastodon, megatherium, taxodon, etc.).

Key Difficulties on Darwin's Theory

Difficulty	*Evidence to Resolve*
1. Lack of interminable numbers of intermediate forms that should exist	Natural selection process exterminates intermediate forms; no right to expect to see *directly* connected forms, only links between organism and some extinct and supplanted form.
2. Lack of fossil remains of interminable number of intermediate forms; sudden appearance of whole groups of species; lack of fossil remains below Silurian system	Fossilization process rare in nature; only small portion of world has been geologically explored; great imperfection of the geological record; only organic beings of certain classes can be preserved in a fossil condition.
3. Complex and useless organs	This difficulty not real if we admit: gradations in the perfection of each organ exist now or could have existed; all organs and instincts are variable to some extent; there is a struggle for existence leading to preservation of each profitable deviation, though in the course of time they may become useless; the truth of the above propositions, Darwin feels, is "indisputable."
4. Instincts	Same reasons as above.
5. Universal sterility of species when first crossed; universal fertility of varieties	Darwin contention that this is not so: cites exceptions.
6. Wide geographical diffusion of some species with no apparent means of dispersal	During long period of time always good chance for wide migration by many means; we are ignorant of past climatic and geographic changes (Darwin's glacial period theory); profoundly ignorant of many occasional means of transport.

Evidence Adduced in Support of Darwin's Theory

Facts of Nature	Darwin's View of a Natural Selection vs. Creationist Explanation
1. Lack of clear demarcation between species and varieties	Makes sense in terms of natural selection theory where varieties are viewed as incipient species; makes no sense from Creationist view.
2. Classification of groups subordinate to groups	Makes sense in terms of natural selection, where large groups increase in size and decrease in character with attendant extinction; "utterly inexplicable" by Creationist view.
3. *"Natura non facit saltum"* (nature makes no leaps)	Natural selection can act only by accumulating slight, successive, favorable variations, and hence supports *natura non facit saltum* idea; why this should be so in Creationist view "no man can explain."
4. Cases of lack of absolute perfection in organisms—e.g., upland geese, which never swim, having webbed feet	Makes sense in terms of natural selection which adapts inhabitants of each country only in relation to the degree of perfection of its inhabitants; meaningless from Creationist point of view.
5. Reversion to long lost characters	Explained by natural selection in sense that characters are those of descendants; "inexplicable" by theory of Creation.
6. Specific characters more variable, generic characters less variable in all species.	Makes no sense in Creationist view; makes sense in natural selection theory, where species are only well-marked varieties and where generic characters are those inherited without change for an enormous period of time.
7. Parts developed in very unusual manner highly susceptible to variation	Inexplicable by theory of creation; by natural selection theory, part has undergone considerable variation and will continue to do so.

	Darwin's View of a
Facts of Nature	*Natural Selection vs.* *Creationist Explanation*

8. The geological record, where new species have come on slowly, where extinction is evident, where species do not reappear once the chain of ordinary generation has been broken, where extinct species are closely related to living forms in closely related geographic areas

Makes sense with natural selection theory where organisms are related in a genealogical way; does not make sense from Cretionist point of view.

9. The leading facts in the geographical distribution of organisms: parallelism in the distribution of organic beings throughout space, and in their geological succession throughout time; the close connection of inhabitants of the same continents under the most diverse geographical and environmental conditions; the relation of oceanic to nearest continental productions; the fact that species which cannot easily cross wide spaces of ocean do not inhabit oceanic islands; the existence of closely allied or representative species

Makes sense with idea of migration and with subsequent modification because the real affinities of all organic beings are due to inheritance or community of descent; "utterly inexplicable" in Creationist view.

Facts of Nature	*Darwin's View of a Natural Selection vs. Creationist Explanation*
10. Morphological evidence—i.e., bones being the same in the hand of a man, wing of a bat, etc.	Makes sense from natural selection point of view, where all organisms are related; inexplicable in Creationist position.
11. Embryonic state of mammals, birds, reptiles, and fishes alike	Makes sense from natural selection point of view, where successive variations do not always supervene at early age; makes no sense on Creationist view.
12. Rudimentary organs	Makes sense from natural selection point of view, where a once important organ has atrophied from disuse; from Creationist point of view, why would useless characteristics be produced?

Notes

Chapter 1

1. Charles Darwin, *The Autobiography of Charles Darwin, 1809–1882, With Original Omissions Restored*, pp. 119–20. There is controversy concerning the exact role that artificial selection played in Darwin's discovery of natural selection. See Michael Ruse, "Charles Darwin and Artificial Selection," and Sandra Herbert, "Darwin, Malthus, and Selection." Herbert maintains that Darwin's discovery of natural selection made the domestic analogy more meaningful to him, while Ruse contends that it is difficult to discount the numerous domestic breeding publications read by Darwin before he read Malthus. See also Camille Limoges, *La selection naturelle.*

2. *The Autobiography of Charles Darwin*, p. 120.

3. For a discussion of these issues, see Peter J. Vorzimmer, "Darwin, Malthus, and the Theory of Natural Selection." Vorzimmer sides with the old view (p. 542): "Without doubt . . . the great watershed in the development of Darwin's evolutionary theory came with his reading of Malthus." See also Herbert, "Darwin, Malthus, and Selection," Robert M. Young, "Malthus and the Evolutionists: The Common Context of Biological and Social Theory," and Gavin de Beer, "The Origins of Darwin's Ideas on Evolution and Natural Selection."

4. Barry G. Gale, "Darwin and the Concept of a Struggle for Existence: A Study in the Extrascientific Origins of Scientific Ideas."

5. See, for example, Gerhard Wichler, *Charles Darwin: The Founder of the Theory of Evolution and Natural Selection;* Gavin de Beer, *Charles Darwin: Evolution by Natural Selection;* Vorzimmer, "Darwin, Malthus, and the Theory of Natural Selection." De Beer writes: "By the end of the year 1838, then, Darwin had discovered the most important principle in biology, comparable in scope to Newton's discovery of gravitation" (p. 106). Vorzimmer writes: " . . . with but the one notable exception of 'divergence,' from 1838 onwards Darwin was able to work with a clear formulation of his theory of natural selection" (p. 542).

Cf. also Ernst Mayr, "Darwin and Natural Selection," and Silvan S. Schweber, "The Origin of the 'Origin' Revisited." Schweber's thesis is that "by August 1838

171

Darwin had indeed apprehended the essential features of the evolutionary mechanism" (p. 231).

6. Cf. Michael T. Ghiselin, *The Triumph of the Darwinian Method*, though Ghiselin's work is more philosophy of science and intellectual history than biography.

7. See Howard E. Gruber and Paul H. Barrett, *Darwin on Man: A Psychological Study of Scientific Creativity*, pp.35–45, for a discussion of Darwin's "long delay."

8. There are numerous works on Darwin written from a psychoanalytic or psychological perspective in which Darwin's psychological makeup is carefully dissected. See, for example, James A. Brussel, "The Nature of the Naturalist's Unnatural Illness: A Study of Charles Darwin"; Phyllis Greenacre, *The Quest for the Father: A Study of the Darwin-Butler Controversy, as a Contribution to the Understanding of the Creative Individual;* Dr. Douglas Hubble, "Darwin's Illness," "The Autobiography of Charles Darwin," "The Life of the Shawl," and "Charles Darwin and Psychotherapy"; Dr. Rankine Good, "The Life of the Shawl," "The Origin of 'The Origin': A Psychological Approach"; Imre Hermann, "Charles Darwin"; Edward J. Kempf, "Charles Darwin—The Affective Sources of His Inspiration and Anxiety Neurosis," reprinted in Kempf, *Psychopathology.*

9. See, for example, Gruber and Barrett, *Darwin on Man*, pp. 35–45.

Chapter 2

1. *Darwin on Man*, see especially Chapter 2, "The Threat of Persecution," pp.35–45. Gruber maintains that Darwin probably knew of Galileo's renunciation of his doctrine of the earth's motion before the Inquisition. The incident is mentioned in William Whewell's *History of the Inductive Sciences*, which Darwin apparently read. Whewell was a Professor at Cambridge during Darwin's student days there, and Darwin knew him. Darwin also must have known of the plight of the French proponent of the new geology, Georges-Louis Leclerc, Comte de Buffon, who in the middle of the eighteenth century was forced by the Sorbonne to renounce all of his views regarding the earth's formation which contradicted Scripture and which were "contrary to the narration of Moses." The story is discussed in vol. I of Lyell's *Principles of Geology*, which we know Darwin studied carefully. Another incident which might have been instructive struck much closer to Darwin personally, Gruber continues. While at Edinburgh University (1825–27), Darwin was a member of the Plinian Society, a student/faculty group which met weekly to discuss matters related to the natural sciences. At a March 27, 1827 meeting, a Mr. Browne read a paper which clearly avowed a materialist position. Apparently some rancor ensued, and the section of the minutes recording the substance of Browne's paper was expunged. The Secretary of the Society went so far as to expunge in earlier minutes even Mr. Browne's *intention* of reading his paper. Gruber sees this incident as "one of Darwin's early exposures to a materialist philosophy of mind and a strong antagonistic reaction to it." Also, Gruber points to several passages in Darwin's Trans-

mutation and M and N notebooks in which Darwin expresses materialist views. In two of the passages he expresses concern regarding the negative reactions these views might arouse. Lastly, Gruber describes with what care a contemporary of Darwin's, the Edinburgh publisher and naturalist, Robert Chambers, strove to keep secret his identity as author of *Vestiges of the Natural History of Creation* (1844), a semipopular scientific book in which (among other things) a theory of evolution is espoused. Chambers's motives for his secrecy were to avoid "bitter and probably painful personal disputes," and to avoid hurting his business interests. The work did cause enormous controversy and bitterness. One of its most ferocious critics was Adam Sedgwick, Professor of Geology at Cambridge, and a former teacher and associate of Darwin's.

2. Cf. Susan F. Cannon, "The Whewell-Darwin Controversy," p. 383.

3. There is also the possibility, suggested to me by Dr. Susan Cannon, that Darwin feared "prejudice" in John Herschel's meaning of the prejudices of opinion and sense that scientists must rid themselves of before attaining truth. We know that Darwin read Herschel's *Preliminary Discourse on the Study of Natural Philosophy* while at Cambridge and was deeply impressed by it. Whether Darwin was impressed by this particular aspect of Herschel's work, it is difficult to say. For Herschel's discussion of prejudices of opinion and sense, see *Preliminary Discourse*, pp.79–84.

4. Charles Darwin, *The Life and Letters of Charles Darwin*, I, p. 271. Hereafter referred to as *Life and Letters*.

5. Ibid., I, p. 393.

6. Ibid., I. p. 394. Date not certain. Probably 1845.

7. Compare Nora Barlow, ed., *Darwin and Henslow: The Growth of an Idea*. As Nora Barlow wrote (p. 8): "When Charles left England in the last days of 1831, he could claim little more than the rank of amateur geologist and naturalist: he returned in five years, a scientist who could command the attention of the great men of the day." I think Barlow overstates the case here.

8. *The Autobiography of Charles Darwin*, pp. 22–28, 43–46.

9. Ibid., p. 44.

10. See J. H. Ashworth, "Charles Darwin as a Student in Edinburgh," p. 98.

11. *The Autobiography of Charles Darwin*, p. 52.

12. Robert Darwin was a forceful person who often lectured to his children. Perhaps Charles had had enough.

With regard to the lectures at Edinburgh, Ashworth has compared Darwin's adverse opinion of the teaching abilities of Monro, Duncan, and Jameson with the views of several gifted Edinburgh students of about the same period. Ashworth has concluded that although all students seem to agree with regard to Monro's deficiencies, on Duncan's and Jameson's abilities opinion differs. These two teachers were thought by the other students to have provided interesting lectures. See "Charles Darwin as a Student at Edinburgh," pp. 99–101.

13. Ibid., pp. 102–3. To mention but a few of the topics of papers presented at the Society: the alleged oviposition of the cuckoo in the nests of other birds, extrauterine gestation, the anatomy of expression, the mode of obtaining bromine from soap-boilers' waste, on the capture of whales on the coast of the Shetlands. Ibid.

14. *The Autobiography of Charles Darwin*, p. 49.

15. MacGillivray later published *A History of British Birds, Indigenous and Migratory, Including their Organization, Habits, and Relations*. MacGillivray was also interested in geological and zoological subjects.

An attempt has been made to identify the black taxidermist who instructed Darwin. See R. B. Freeman, "Darwin's Negro Bird-Stuffer."

16. "Charles Darwin as a Student in Edinburgh," pp. 106–9.

17. Ibid.

18. *The Autobiography of Charles Darwin*, pp. 47, 50.

19. See P. Helveg Jespersen, "Charles Darwin and Dr. Grant." Darwin later had a priority problem with Wallace. See Chapter 8.

20. *The Autobiography of Charles Darwin*, pp. 54–55.

21. Ibid., pp. 58–62.

22. *Life and Letters*, I, p. 141. Francis Darwin's recollections.

23. *The Autobiography of Charles Darwin*, p. 60.

24. Ibid.

25. *Life and Letters*, I, p. 150.

26. Ibid. The entomologist Frederick Hope, who was later to help Darwin with the *Beagle* specimens. Hope was founder of the Chair of Zoology at Oxford.

27. Ibid., I, p. 154. July 18, 1829.

28. Ibid., I, p. 156.

29. Ibid., I, p. 157. November 5, 1830.

30. *The Autobiography of Charles Darwin*, p. 62.

31. Ibid., p. 64.

32. Ibid.

33. Ibid. See Barlow, *Darwin and Henslow*, pp. 4–7 for a description of Henslow's activities and stature at Cambridge during this time.

34. *The Autobiography of Charles Darwin*, p. 64. For a recent study on Henslow which emphasizes his popular science education efforts in mid-Victorian England, see Jean Russell-Gebbett, *Henslow of Hitcham*.

35. *The Autobiography of Charles Darwin*, pp. 67–68. For Humboldt's influence on Darwin, see Frank N. Egerton, "Humboldt, Darwin and Population."

36. Paul H. Barrett, "The Sedgwick-Darwin Geologic Tour of North Wales," pp. 146, 151.

37. *The Autobiogaphy of Charles Darwin*, p. 68. Emphasis mine.

38. Ibid., p. 70. Emphasis mine.

39. *Life and Letters*, I, p. 164.

40. Barlow, *Darwin and Henslow*, p. 25.

41. *The Autobiography of Charles Darwin*, p. 71. Emphasis mine.

42. Barlow, *Darwin and Henslow*, p. 28.

43. Ibid., p. 30. Emphasis Henslow's. See also Gavin de Beer, "The Darwin Letters at Shrewsbury School," for an assessment of the state of Darwin's knowledge at the time. Cf. also Harold Fruchtbaum, *Times Literary Supplement*, October 5, 1967, p. 938. Fruchtbaum maintains that when Darwin sailed on the *Beagle* he was "one of the best-trained and most experienced naturalists in England." De Beer can see no evidence for this statement, nor can I.

44. *The Autobiography of Charles Darwin*, p. 76. We now associate the voy-

age of the *Beagle* with Darwin and his naturalist experiences in South America, the Galapagos Islands, etc. Yet, during that time the voyage's primary purpose was hydrographic.

The years between 1830 and 1880, George Basalla tells us, saw Britain lead the world in hydrography. Britain's greatest hydrographer during the first half of that period was Captain Frances Beaufort—a friend of Peacock's, from whom, apparently, Peacock first learned of the *Beagle* opportunity. Beaufort was head of the Royal Navy's Hydrographic Department. In the course of his tenure as head of the Department, he sent out some 170 major surveying expeditions. The *Beagle* survey was one of the first of these expeditions.

The voyage proved a great boon to FitzRoy, establishing him as an hydrographer of the first rank. It was his second such expedition. As product of the voyage, FitzRoy furnished the Admiralty with 82 coastal sheets, 80 plans of harbors, and 40 views covering the southern part of the South American continent. In addition, the *Beagle* was supplied with 22 chronometers so that FitzRoy could establish a chain of meridian distances through the Pacific, Indian, and Atlantic Oceans. For his efforts during the voyage, FitzRoy was awarded the Royal Geographical Society's 1837 medal.

The *Beagle* voyage, then, was really FitzRoy's voyage, not Darwin's. See George Basalla, "The Voyage of the *Beagle* without Darwin."

45. Jacob W. Gruber, "Who was the *Beagle*'s Naturalist?" See Harold Burstyn, "If Darwin Wasn't the *Beagle*'s Naturalist, Why was He on Board?" for the view of Darwin as gentlemanly companion.

46. See Darwin's letter to Henslow (September 1831) in *Darwin and Henslow*, p. 38. For Darwin's relationship with FitzRoy, see Nora Barlow, "Robert FitzRoy and Charles Darwin."

47. Barlow, *Darwin and Henslow*, p. 29.

48. *Life and Letters*,I, p. 182.

49. Ibid., p. 187.

50. Ibid., p. 202.

51. Barlow, *Darwin and Henslow*, p. 38.

52. Charles Darwin, *Journal of Researches into the Geology and Natural History of the Various Countries Visited by H.M.S. Beagle.*

53. *Life and Letters*, I, p. 220. June 2, 1833.

54. Barlow, *Darwin and Henslow*, p. 43.

55. Ibid., p. 50.

56. Ibid., p. 56.

57. Ibid., pp. 58–59. August 15.

58. Ibid., p. 61. November 24.

59. Ibid., pp. 65–67. January 15. Emphasis Henslow's.

60. Ibid., p. 74. April 11.

61. Ibid., p. 79. August 31. Emphasis Henslow's.

62. Ibid.

63. Ibid., p. 84.

64. Ibid., p. 89. July 22. Emphasis Henslow's.

65. Ibid., p. 92. July 24.

66. Ibid., p. 113.

67. Darwin, *Journal of Researches*, p. ix. For more on Darwin and his activities during the voyage of the *Beagle*, see R. D. Keynes, "Darwin and the *Beagle*."

68. *The Autobiography of Charles Darwin*, p. 81.

69. For an analysis of Darwin's interests in geology during the voyage, see Howard E. Gruber and Valmai Gruber, "The Eye of Reason: Darwin's Development During the *Beagle* Voyage."

70. *Life and Letters*, I, p. 203.

71. Ibid., I, pp. 206–7.

72. Barlow, *Darwin and Henslow*, pp. 53, 56–57. May 18.

73. *Life and Letters*, I, p. 227. July 23.

74. Ibid., I, pp. 234–35. July 1835.

75. Barlow, *Darwin and Henslow*, p. 115. July 9.

76. *The Autobiography of Charles Darwin*, pp. 78–79.

77. Ibid., p. 77.

78. Barlow, *Darwin and Henslow*, p. 85.

79. Ibid., p. 93. July 24.

80. *Life and Letters*, I, p. 237.

81. Barlow, *Darwin and Henslow*, pp. 62–63.

82. Ibid., p. 83.

83. *The Autobiography of Charles Darwin*, pp. 77–78.

84. Barlow, *Darwin and Henslow*, pp. 123–24. David Don was professor of botany, King's College, London, and librarian to the Linnean Society.

85. For an appreciation of the hectic nature of the period following Darwin's return from the *Beagle* voyage, see Schweber, "The Origin of the 'Origin' Revisited," pp. 229–30.

86. *Darwin and Henslow*, p. 119. October 30, 1836.

87. Ibid., p. 122.

88. *Life and Letters*, I, p. 248. November 6.

89. Ibid., I, 252–53. April 10, 1837.

90. It finally appeared in five volumes as *Zoology of the Voyage of H.M.S. Beagle*, edited and superintended by Charles Darwin.

91. *Life and Letters*, I, p. 250. March 1837 letter to Fox.

92. Getting help from others would later be a way of life for Darwin. See Chapters 4 and 5.

93. Barlow, *Darwin and Henslow*, p. 128.

94. Ibid., pp. 131–32. July 14.

95. Ibid., p. 133. Emphasis Darwin's.

96. Ibid., p. 131. May 28.

97. Ibid., pp. 138–40. Emphasis Darwin's.

98. *Life and Letters*, I, p. 268. September 13. Emphasis Darwin's.

99. Charles Darwin, *Darwin's Notebooks on Transmutation of Species*. Hereafter referred to as Transmutation Notebooks, I through IV. (In references to the Transmutation Notebooks, Roman numerals refer to the notebook number, Arabic numerals to the page number of the original notebook, *not* to the page number in de Beer's edition.)

Many other examples can be cited. See Transmutation Notebooks, I, 48, 112,

158, 168, 180, 193, 220, 230, 251, 263; II, 21, 46, 61, 67, 73, 90, 99, 166, 198, 204, 219, 220, 224, 233, 243, 265, 267; Excised pages—II, 42, 49, 184; III, 151; IV, 9, 119. Emphasis Darwin's.

See also Paul H. Barrett, "A Transcription of Darwin's First Notebook on 'Transmutation of Species.'"

100. For a discussion of the state of Darwin's knowledge following the *Beagle* voyage, cf. Schweber, "The Origin of the 'Origin' Revisited," p. 310; Frank N. Egerton, "Refutation and Conjecture: Darwin's Response to Sedgwick's Attack on Chambers"; and Gavin de Beer, ed. *Evolution by Natural Selection*, pp. 3–4. See also George Grinnell, "The Rise and Fall of Darwin's First Theory of Transmutation."

Chapter 3

1. Gavin de Beer, ed. "Darwin Journal," pp. 1–21. See Sydney Smith, "The Origin of the *Origin* as Discerned from Charles Darwin's Notebooks and His Annotations in the Books He Read between 1837 and 1842," pp. 391–401, especially p. 393, for a discussion of Darwin's *Diary.*

2. *Life and Letters*, I, pp. 377–79. The botanist Joseph Dalton Hooker became Darwin's close friend. See Chapter 4.

3. Ibid., I, p. 391. Another reason he continued working on Cirripedes was because of the advice given him by Hooker. See Chapter 4.

For Darwin and Cirripedes, cf. Thaddeus J. Trenn, "Charles Darwin, Fossil Cirripedes, and Robert Fitch: Presenting Sixteen Hitherto Unpublished Darwin Letters of 1849 to 1851," and Sydney Smith, "The Darwin Collection at Cambridge With One Example of Its Use: Charles Darwin and Cirripedes."

4. *Life and Letters*, I, p. 315.

5. Ibid., I, p. 314.

6. See de Beer, "Darwin's Journal," p. 13.

7. Cf. Francis Darwin's "Reminiscenses," in *Life and Letters*, I, pp. 87–136. In addition to being a thorough worker, Darwin was a bold and imaginative theorizer. The two characteristics often seemed to be in struggle.

Darwin's work on Cirripedes appeared as *A Monograph of the Fossil Lepadidae, or Pedunculated Cirripedes of Great Britain; A Monograph of the Sub--Class Cirripedia, with figures of all Species. The Lepadidae, or Pedunculated Cirripedes; A Monograph of the Sub--Class Cirripedia. The Balanidae...The Verruciade;* and *A Monograph of the Fossil Balanidae and Verruciade of Great Britain.*

8. For Darwin's relationship to Lyell, see Michael Bartholomew, "The Non-Progress of Non-Progression: Two Responses to Lyell's Doctrine"; Joe O. Burchfield, "Darwin and the Dilemma of Geologic Time"; Leonard G. Wilson, *Charles Lyell: The Years to 1841;* Martin J. S. Rudwick, "The Strategy of Lyell's *Principles of Geology;*" Sir Edward Bailey, *Charles Lyell;* William Coleman, "Lyell and the 'Reality' " of Species: 1830–1833"; Susan F. Cannon, "The Unformitarian-Catastrophist Debate," "Charles Lyell, Radical Actualism, and Theory," and "The Whewell-Darwin Controversy"; Bert James Loewenberg, "The Mosaic of

Darwinian Thought"; Gertrude Himmelfarb, *Darwin and the Darwinian Revolution;* Loren Eiseley, *Darwin's Century;* William Irvine, *Apes, Angels, and Victorians;* Charles C. Gillispie, *Genesis and Geology.*

9. Charles Lyell, *The Principles of Geology.*

10. Ibid., II, pp. 3–17.

11. For more on Lyell's theories, see Martin J. S. Rudwick, "Charles Lyell, F.R.S. (1797–1875) and his London Lectures on Geology, 1832–1833," and Susan F. Cannon, "Charles Lyell is Permitted to Speak for Himself: An Abstract," in *Toward a History of Geology,* ed. Cecil J. Schneer.

12. See Himmelfarb, *Darwin and the Darwinian Revolution,* pp. 85–89, and Cannon, "The Uniformitarian-Catastrophist Debate," pp. 38–41.

13. Darwin tried to argue in a similar way regarding his natural selection theory. See Chapters 6 and 7.

14. Adam Sedgwick, "Address to the Geological Society, Delivered on the Evening of the Anniversary, February 18, 1831." Cf. Susan F. Cannon, "The Problem of Miracles in the 1830's." Cannon maintains that in the 1830s Christians tried to define the miraculous so "that it could be related rather than opposed to natural science proper." (p. 5.)

15. Whewell's review of Lyell's *Principles* in *The British Critic Quarterly Theological Review.* For more on Uniformitarian and Progressionist views, see Leonard G. Wilson, "Geology on the Eve of Charles Lyell's First Visit to America, 1841"; "The Intellectual Background to Charles Lyell's *Principles of Geology,* 1830–1833"; Martin J. S. Rudwick, "Historical Analogies in the Geological Work of Charles Lyell"; Peter J. Bowler, *Fossils and Progress: Paleontology and the Idea of Progressive Evolution in the Nineteenth Century;* Susan F. Cannon, "The Impact of Uniformitarianism: Two letters from John Herschel to Charles Lyell, 1836–37"; R. Hooykaas, "The Principle of Uniformity in Geology, Biology, and Theology," and *Natural Law and Divine Miracle. A Historical-Critical Study on the Principle of Uniformity in Geology.*

16. Cf. de Beer, *Charles Darwin,* p. 57.

17. *Darwin and Henslow,* pp. 118, 122. Emphasis Darwin's.

18. *Life and Letters,* I, p. 253.

19. Ibid., I, p. 251. Lyell probably first met Darwin at Henslow's in June 1831 before the *Beagle* voyage. See Sydney Smith, "The Origin of the 'Origin'," p. 397.

20. Kinnordy mss. Quoted in Wilson, *Charles Lyell,* p. 441.

21. Darwin and Seward, eds., *More Letters of Charles Darwin: A Record of His Work in a Series of Hitherto Unpublished Letters.* Hereafter referred to as *More Letters.*

22. *Life and Letters,* I, p. 307.

23. Ibid., I, p. 313. October 8.

24. October 26, 1836. Owen mss. Quoted in Wilson, *Charles Lyell,* p. 434. For a history and analysis of the Geological Society of London, which came to rival the Royal Society in esteem, see Martin J. S. Rudwick, "The Foundation of the Geological Society of London: Its Scheme for Co-operative Research and Its Struggle for Independence." See also J. B. Morrell, "London Institutions and Lyell's Career: 1820–1841."

25. Charles Darwin, "Geological Notes Made During A Survey of the East

and West Coasts of South America, in the Years 1832, 1833, 1834, and 1835, with an Account of A Transverse Section of the Cordilleras of the Andes between Valparaiso and Mendoza."

26. Ibid., II, p. 367.

27. Ibid., II, p. 446–49.

28. See Wilson, *Charles Lyell*, pp. 434–35.

29. *Proceedings of the Geological Society of London*, II:504–6.

30. Ibid., II, p. 543.

31. Ibid., II, p. 511.

32. *Principles of Geology*, II, pp. 283–86.

33. "On certain areas of elevation and subsidence in the Pacific and Indian oceans as deduced from the study of Coral Formations."

34. De Beer, *Charles Darwin*, p. 71.

35. Lyell, ed., *The Life, Letters and Journals of Charles Lyell*, p. 12. Hereafter referred to as *Life of Lyell*.

36. *Life and Letters*, I, p. 294. 1837 letter.

37. Cf. D. R. Stoddart, "Darwin, Lyell, and the Geological Significance of Coral Reefs," and C. M. Yonge, "Darwin and Coral Reefs," in Barnett, *A Century of Darwin*, pp. 245–67.

Lyell was also impressed with several other of Darwin's geological works, including: Darwin's study of the parallel roads of Glen Roy, which Darwin explained, again in good Uniformitarian fashion, as the elevated shores of an ancient sea ("Observations on the Parallel Roads of Glen Roy, and other part of Lochaber in Scotland, with an attempt to prove that they are of marine origin"; see also Martin J. S. Rudwick, "Darwin and Glen Roy: A 'Great Failure' in Scientific Method?"); Darwin's description and analysis of South American earthquakes and volcanoes, which originally appeared as the second volume of Darwin's geology of the voyage of the *Beagle* (see *Geological Observations on Volcanic Islands and Parts of South America*); and Darwin's newly completed (1837) *Journal of Researches* (which, to Darwin's delight, Lyell cited often in a new, 1838 edition of the *Principles*). "You will see I am in a fit of enthusiasm, and good cause I have to be," Darwin wrote Lyell on August 9, 1838, "when I find you have made such infinitely more use of my Journal than I could have anticipated." (*Life and Letters*, I, p. 263.)

38. Ibid., I, p. 267.

39. *More Letters*, II, p. 117.

40. *Life and Letters*, I, p. 310. August 25, 1845. A. R. Wallace seemed similarly impressed. See *The Darwin-Wallace Celebration held on Thursday, 1st July, 1908, by the Linnean Society of London*, p. 111.

41. Quoted in *Life and Letters*, I, p. 306.

42. Ibid.

43. Ibid., I, p. 266. September 13, 1838.

44. *More Letters*, I, p. 173.

45. *Life of Lyell*, I, p. 475. December 26, 1836.

46. Litchfield, ed., *Emma Darwin, A Century of Family Letters, 1792–1896*, II, p. 14. Hereafter referred to as *Emma Darwin*. November 23, 1838, Darwin to Emma.

47. *Life and Letters*, I, p. 268. September 13, 1838.

48. *Life of Lyell*, II, p. 45. September 8, 1838.
49. *Life and Letters*, I, p. 264. August 9, 1838.
50. *Life of Lyell*, II, p. 46.
51. *Life and Letters*, I, p. 264.
52. *Life of Lyell*, I, p. 475. December 26, 1836.
53. Kinnordy mss. Quoted in Wilson, *Charles Lyell*, p. 459. September 1838 (?).
54. *Life of Lyell*, I, pp. 474–75.
55. "On the connexion of certain volcanic phenomena." Quoted in Wilson, *Charles Lyell*, p. 455.
56. Ibid.
57. Quoted ibid., pp. 455–56.
58. *Life of Lyell*, II, p. 44.
59. *Life and Letters*, I, p. 267. Emphasis Darwin's.
60. Emma to Elizabeth Wedgwood, *Emma Darwin*, II, pp. 23–24.
61. *Life and Letters*, I, p. 260.
62. Ibid., I, p. 271.
63. Ibid., I, p. 272.
64. Ibid., I, p. 243. The literature on Darwin's illness has become voluminous. For three recent treatments, see Ralph Colp, Jr., *To be an Invalid: The Illnesses of Charles Darwin*; George Pickering, *Creative Malady: Illness in the Lives and Minds of Charles Darwin et al.*, Chapters 3 and 4, pp. 34–70; John H. Winslow, *Darwin's Victorian Malady: Evidence of Its Medically Induced Origin*. I think Colp's treatment is especially good in the sense of providing a powerful feeling of the gravity of Darwin's illness and the place that it occupied in his life. For other studies of Darwin's health, see Michael Kelly, "Darwin Really Was Sick"; Hyman J. Roberts, "Reflections on Darwin's Illness"; A. W. Woodruff, "The Impact of Darwin's Voyage to South America on His Work and Health," and "Darwin's Health in Relation to His Voyage to South America"; W. D. Foster, "A Contribution to the Problem of Darwin's Ill-health"; P. B. Medewar, "Darwin's Illness"; Lawrence A. Kohn, "Charles Darwin's Chronic Ill Health"; and S. Adler, "Darwin's Illness."
   For a list of works treating Darwin's illness from a psychological or psychoanalytical point of view, see Chapter 1, note 8.
65. *Life and Letters*, I, pp. 87–136.
66. Ibid., I, pp. 106–7. For other descriptions of Darwin's life at Down, see Leonard Darwin, "Memories of Down House," and Sir Hedley Atkins, *Down: The Home of the Darwins: The Story of a House and Its People.* Sir Hedley lives at Down House and is Curator of the Down House Estate. For some of the changes in Darwin's personality over the Down years, see Donald Fleming, "Charles Darwin, The Anaesthetic Man." Fleming feels that over the years Darwin experienced an atrophy of his aesthetic senses. Darwin admits as much himself.
67. *Life and Letters*, I, p. 345. As the years passed, Darwin's love for geology continued, though he realized that active work in the field was out of the question. In a letter written on February 18, 1854 to Lyell, we get an indication of his longing and frustration: "It really makes me quite envious to think of your

clambering up and down those steep valleys. . . . I do quite envy you. How I should like to be with you, and speculate on the deep and narrow valleys." (Ibid., I, p. 357.)

68. Ibid., I, p. 303.
69. *More Letters*, II, pp. 116–117.
70. *Life and Letters*, I, p. 306.
71. Ibid., I, p. 343. Emphasis Darwin's. April 19, 1849.
72. *The Autobiography of Charles Darwin*, p. 45. Emphasis Darwin's.
73. Ibid., p. 51.
74. Ibid., p. 63.
75. Ibid., pp. 80–81.
76. See Chapter 2.
77. *The Autobiography of Charles Darwin*, p. 82.
78. *Life and Letters*, I, p. 248.
79. Ibid., I, p. 259.
80. Ibid., I, p. 251. In 1849, Darwin read the American geologist and mineralogist James Dwight Dana's study of coral reefs, which confirmed Darwin's coral theory of some ten years earlier. An ecstatic Darwin wrote to Lyell on December 4, 1849: "To begin with a modest speech, *I am astonished at my own accuracy!!* . . . Dana talks of agreeing with my theory *in most points;* I can find out not one in which he differs. Considering how infinitely more he saw of Coral Reefs than I did, this is wonderfully satisfactory to me. . . . There now, my vanity is pretty well satisfied. . . . " (*Life and Letters*, I, p. 342.) Emphasis Darwin's.
81. *The Autobiography of Charles Darwin*, Note 3, pp. 231–32.

Chapter 4

1. Biographical studies of Hooker are surprisingly few in number. One of the earliest is still the best—*The Life and Letters of Sir Joseph Dalton Hooker*, ed. Leonard Huxley, hereafter referred to as *Life of Hooker*. Sir E. Ray Lankester's 1918 review of *Life of Hooker* is insightful ("A Great Naturalist—Sir Joseph Hooker."). More recently, William B. Turrill's *Pioneer Plant Geography: The Phytogeographical Researches of Sir Joseph Hooker* concentrates on Hooker's technical contributions to botany, with very little mention of Hooker's work with Darwin, while Mea Allan's *The Hookers of Kew, 1785–1911* is a semi-popular account of William and Joseph Dalton Hooker's efforts at Kew. Allan does treat Hooker's work with Darwin in a short chapter. A recent article, Janet Browne's "The Charles Darwin-Joseph Hooker Correspondence: An Analysis of Manuscript Sources and Their Use in Biography," looks at the interaction of geology and botany in the Hooker-Darwin relationship and suggests some possible discrepancies in the dating of the first few letters between the two men. Mea Allan has recently published a work on Darwin and botany: *Darwin and His Flowers: The Key to Natural Selection.*

For more on Hooker and Kew Gardens, see Robert H. Scott, "The History of Kew Observatory," and R. G. C. Desmond, "The Hookers and the Development of the Royal Botanic Gardens, Kew."

2. *Life and Letters*, I, p. 380.

3. This seems to support Jacob Gruber's point about a ship's surgeon serving as naturalist. See Chapter 2.

4. Ibid., I, p. 381.

5. Joseph Dalton Hooker, *The Botany of the Antarctic Voyage of H.M. Discovery Ships Erebus and Terror in the Years 1839–1843, Under the Command of Captain Sir James Clark Ross.*

6. He was later to be president of that prestigious body for five years, 1873 through 1878.

7. Joseph Dalton Hooker, *Himalayan Journals; or, Notes of a Naturalist in Bengal, The Sikkim and Nepal Himalayas, The Khasia Mountains, &c.*

8. Joseph Dalton Hooker and Thomas Thomson, *Introductory Essay to the Flora Indica.* The idea for further volumes was dropped.

9. *More Letters*, II, p. 242.

10. *Life and Letters*, I, pp. 382–83.

11. Ibid., I, pp. 381–82.,

12. *More Letters*, I, p. 39.

13. Ibid., II, p. 120. To Lyell, September 1844.

14. *Life and Letters*, I, p. 385.

15. Ibid., I, p. 384.

16. As we will remember, in Darwin's December 1843 letter to Hooker, quoted above, he spoke of the "botanical ignoramuses like myself" that needed help and asked Hooker to name, based on his review of the collections, which genera were American, which European "for the sake of the ignoramuses. . . . " And several years later, in September 1846, Darwin referred to himself in a letter to Hooker as a "man who hardly knows a Daisy from a Dandelion. . . . " (Ibid., I, p. 313.)

17. Ibid., I, p. 304. February 10.

18. Quoted in *Life of Hooker*, I, p. 489. For more on geographical distribution and the role it played in the development of Darwin's theories, see R. Alan Richardson, "Biogeography and the Genesis of Darwin's Ideas on Transmutation."

19. For Darwin and geographic isolation, see Frank J. Sulloway, "Geographic Isolation in Darwin's Thinking: The Vicissitudes of A Crucial Idea."

20. Quoted in *Life of Hooker*, I, p. 487. June 19, 1860. In a July 19, 1856 letter to Hooker Darwin wrote: "It is the greatest temptation to me to write *ad infinitum* to you." (Ibid.)

21. *Life and Letters*, I, pp. 494–95. October 13, 1858 letter to Hooker. Other friends did not react so sympathetically. Leonard Jenyns, for example, had Darwin on the defensive in a follow-up letter to one in which Darwin revealed his thoughts about the mutability of species. "With respect to my far distant work on species, I must have expressed myself with singular inaccuracy if I led you to suppose that I meant to say that my conclusions were inevitable . . . ," Darwin cautiously advised Jenyns. "I never expect more than to be able to show that there are two sides to the question of the immutability of species. . . . " (Ibid., I, p. 394. 1845 [?] letter to Jenyns.)

22. *More Letters*, I, p. 40. January 11, 1844 letter to Hooker.

23. Ibid., I, p. 403. March 31, 1844 letter to Hooker.

24. *Life and Letters*, I, p. 391. July 1844 letter to Hooker.

25. *More Letters*, I, p. 57. September 1846 letter to Hooker.

26. Ibid., I, p. 40.

27. Ibid., I, p. 41.

28. *Life and Letters*, I, p. 321. April 17.

29. *More Letters*, I, p. 63.

30. Ibid., I, p. 47.

31. *Life and Letters*, I, p. 416. I have not been able to determine with any reasonable accuracy how many experiments were performed by Hooker for Darwin or what proportion of Darwin's total experiments were performed by Hooker. My guess on the latter question would be about 10 to 15 percent.

32. *More Letters*, I, p. 95.

33. *Life and Letters*, I, p. 503.

34. Ibid., I, p. 323.

35. Ibid., I, p. 461. Emphasis Darwin's.

36. See *Life and Letters*, I, p. 387. Hooker's later recollection. See Chapter 5.

37. Ibid., p. 304.

38. Ibid., I, p. 328. October.

39. Ibid., I, pp. 320–21.

40. *More Letters*, I, p. 47.

41. Ibid., I, pp. 416–17.

42. *Life and Letters*, I, p. 362. September 1854.

43. Ibid., I, p. 325.

44. *More Letters*, I, p. 419.

45. Ibid., I, p. 62.

46. Ibid., I, pp. 412–13.

47. Ibid., I, p. 62.

48. *Life and Letters*, I, p. 391.

49. *More Letters*, I, p. 56.

50. *Life and Letters*, I, p. 520.

51. Ibid., I, p. 325.

52. Ibid., I, p. 327. Many years later, in 1860, Darwin felt that his theory of the origin of coal had been proved correct. As he wrote to Hooker on May 22 of that year: "Lyell tells me that Binney has published in Proceedings of Manchester Society a paper trying to show that Coal plants must have grown in very marine marshes. Do you remember how savage you were long years ago at my broaching such a conjecture?" (*More Letters*, II, pp. 219–20.)

53. "On the Connexion between the Distribution of the Existing Fauna and Flora of the British Isles and the Geological Changes which have affected their area."

54. *Life and Letters*, I, p. 432. June 18, 1856.

55. Ibid., I, p. 431.

56. Ibid., I, p. 438.

57. Ibid., I, p. 440. Forbes's theory of continental extensions is the only instance that I am aware of where Hooker supported a more "speculative" position than Darwin.

58. See de Beer, *Charles Darwin*, pp. 144–47.
59. *Life and Letters*, I, pp. 413–14.
60. Ibid., I, pp. 414–15. April 14, 1855.
61. Quoted in *Life of Hooker*, I, pp. 493–94. No date given. Probably 1856.
62. Quoted ibid., I, p. 494.
63. See de Beer, *Charles Darwin*, p. 146.
64. *Life of Hooker*, I, p. 439. Emphasis Hooker's.
65. *More Letters*, I, p. 51. 1845.
66. Ibid., I, p. 47.
67. *Life and Letters*, I, p. 323.
68. *More Letters*, I, p. 65.
69. *Life and Letters*, I, p. 403.
70. Ibid., I, p. 402.
71. *More Letters*, I, p. 95.
72. Ibid., I, p. 105.
73. *Life and Letters*, I, p. 494. October 12, 1858. Emphasis Darwin's.
74. *Life of Hooker*, I, p. 464. Emphasis Hooker's.
75. *Life and Letters*, I, p. 503. Emphasis Darwin's.
76. Ibid., I, pp. 512–13. Emphasis Darwin's.
77. *Life of Hooker*, I, p. 439. Emphasis Hooker's.
78. *Life and Letters*, I, p. 441.
79. Ibid., I, p. 499. Huxley too had difficulty understanding Darwin's theory, even after the *Origin* had appeared and Huxley had become one of Darwin's chief advocates. See Chapter 5.
80. See *Life of Hooker*, I, p. 498.
81. Ibid., I, p. 438.
82. Ibid., I, p. 439.
83. Ibid., I, p. 441. Emphasis Hooker's.
84. *More Letters*, I, pp. 70–71.
85. *Life and Letters*, I, p. 347. October 12, 1849 letter to Hooker.
86. Ibid., I, p. 499. January 20, 1859.
87. *Life of Hooker*, I, p. 455. March 14, 1858.
88. *Life and Letters*, I, p. 465. May 6, 1858.
89. Ibid., I, p. 450. April 1857. Emphasis Darwin's. I think that Darwin was really defending himself here.
90. Ibid., I, pp. 383–84.
91. *More Letters*, I, p. 39.
92. Quoted in *Life of Hooker*, I, p. 490.
93. Quoted Ibid., I, p. 489.
94. Quoted Ibid., I, p. 496.
95. Quoted Ibid., I, p. 496. Most probably December 1857.
96. *Life and Letters*, I, p. 529.
97. Quoted in *Life of Hooker*, I, p. 490.
98. *Life and Letters*, I, p. 391. September 1845.
99. Ibid., I, p. 360.
100. Ibid., I, p. 442. Emphasis Darwin's.
101. Ibid., II, pp. 20–21. Emphasis Darwin's.

102. Quoted in *Life of Hooker*, I, p. 490.

103. Quoted Ibid., I, p. 490.

104. Ibid., I, p. 451. April 11, 1857.

105. Ibid.

106. *Life and Letters*, I, p. 387.

107. *More Letters*, I, p. 47. 1845.

108. Quoted in *Life of Hooker*, I, p. 491. No date given.

109. *More Letters*, II, p. 339.

Chapter 5

1. See Francis Darwin's "Reminiscences," *Life and Letters*, I, p. 104.

2. Ibid., I, pp. 127–28. Darwin, Francis tells us, did not respect books as such. He usually considered them tools to be utilized. He did not bind his books, and he often cut a large book in two for ease of reading. When a book fell apart he unceremoniously put a clip on it to hold it together. Pamphlets he treated with even less respect—he would tear out and throw away all the pages that did not interest him. See ibid., I, p. 127.

For the books that Darwin actually read, see Peter J. Vorzimmer, "The Darwin Reading Notebooks (1838–1860)."

3. As Darwin wrote to Hooker on April 12, 1857: "I have been interested in my 'weed garden,' of 3x2 feet square: I mark each seedling as it appears, and I am astonished at the number that come up, and still more at the number killed by slugs. . . . " (*Life and Letters*, I, p. 449.)

4. As Darwin wrote to Hooker on June 3, 1857: "Out of sixteen kinds of seed sown on my meadow, fifteen have germinated, but now they are perishing at such a rate that I doubt whether more than one will flower." (Ibid., I, p. 457.)

5. See Darwin's letter to Hooker dated June 5, 1855, for a description of these activities. Ibid., I, pp. 418–19.

6. For a description, see Darwin's letter to Fox dated March 19, 1855, and his letter to Hooker dated 1855. Both ibid., I, p. 407.

7. As Darwin wrote to Hooker in June 1855: "I read somewhere that no plant closes its leaves so promptly in darkness, and I want to cover it up daily for half an hour, and see if I can *teach it* to close by itself, or more easily than at first in darkness." (*More Letters*, I, p. 249.) Emphasis Darwin's.

8. The Columbarian and Philoperistera Clubs. It was at these clubs that Darwin learned that enthusiasm was the key to good pigeon breeding. He wrote humorously to T. H. Huxley that pigeon breeding had become so specialized an art that a Mr. Eaton had written a whole treatise on the Almond Tumbler, which was a sub-variety of the short-faced Tumbler, which in turn was a variety of the rock pigeon. See *Life and Letters*, I, p. 411. For Darwin's work on pigeons and domestication in general, see James A. Secord, "Nature's Fancy: Charles Darwin and the Breeding of Pigeons," and David Kohn, "Theories to Work By: Rejected Theories, Reproduction, and Darwin's Path to Natural Selection."

There are indications throughout the Transmutation and so-called M and N notebooks of Darwin's interests in experiments. (The M and N notebooks are transcribed in Gruber and Barrett, *Darwin on Man*.) For example, we learn that Darwin thought of several ideas for psychological experiments; he carefully observed both his own and other children's behavior, especially with regard to their expression of emotions; he seems to have taken every opportunity possible to experiment with monkeys at the zoo—holding up a mirror to them, making faces, offering and then withdrawing various nuts and recording their reactions. Cf. Transmutation Notebooks, I, 248; II, 212; M Notebook, 89–97, 105–9, 142–57. One of his most interesting experiments, I think, was on the effects of habit. As he wrote in Transmutation Notebook, II, 217: "Case of habit: I kept my tea in right hand for some months, & then when that was finished kept it in left, but I always for a week took of[f] cover of right side though my hand would sometimes vibrate seeing no tea brought back memory—old habit of putting tea in pot, made me go to tea chest almost unconsciously."

9. On the first questionnaire, see Charles Darwin, *Questions About the Breeding of Animals* [1840]; R. B. Freeman and P. J. Gautrey, "Darwin's *Questions about the Breeding of Animals*, with a Note on *Queries about Expression*," and "Charles Darwin's *Queries about Expression*"; Peter J. Vorzimmer, "Darwin's *Questions about the Breeding of Animals* (1839)." See also Roger J. Wood, "J. Robert Bakewell (1725–1795), Pioneer Animal Breeder and His Influence on Charles Darwin," and J. Hammond, "Darwin and Animal Breeding," in *A Century of Darwin*, pp. 85–102.

For the second questionnaire, see Darwin's March 1858 letter to David Forbes in Charles Darwin, "Some Unpublished Letters of Charles Darwin." Hereafter referred to as "Unpublished Letters."

10. *Life and Letters*, I, pp. 387–88.

11. Numerous others are mentioned by Darwin in the *Origin*. The naturalist Edward Blyth was one of these. For a discussion of the correspondence between Darwin and Edward Blyth, see Barbara G. Beddall, "Wallace, Darwin, and Edward Blyth: Further Notes on the Development of Evolution Theory," and "'Notes for Mr. Darwin': Letters to Charles Darwin from Edward Blyth at Calcutta: A Study in the Process of Discovery."

For Edward Blyth's possible influence on Darwin, see Loren Eiseley, "Charles Darwin, Edward Blyth, and the Theory of Natural Selection," and *Darwin and the Mysterious Mr. X: New Light on the Evolutionists*. See also Theodore Dobzhansky, "Blyth, Darwin, and Natural Selection." For a critique of Eiseley's ideas regarding Blyth's influence, see Joel S. Schwartz, "Charles Darwin's Debt to Malthus and Edward Blyth." I think Eiseley's theory is overdrawn. See my comments on other aspects of Eiseley's work in Chapter 8.

12. In a December 16, 1859 letter to Huxley. *More Letters*, I, p. 131.

13. For Lyell's help, cf. *Life and Letters*, I, pp. 426–532.

14. See, for example, Darwin's October 12, 1845 letter to Jenyns. Ibid., I, p. 392.

15. See, for example, Darwin's July 18, 1856 letter to Woodward. *More Letters*, I, p. 96.

16. See Gavin de Beer, ed., "Further Unpublished Letters of Charles Darwin." For Darwin's relationship to John Edward Gray on the subject of barnacles, see A. E. Gunther, "J. E. Gray, Charles Darwin, and the *Cirripedes*, 1846–1851."

17. "Unpublished Letters," pp. 14–27.

18. See, for example, Darwin's July 2, 1855 letter to Henslow. Barlow, *Darwin and Henslow*, pp. 176–78.

19. "Unpublished Letters," pp. 28–32.

20. See *Life and Letters*, I, p. 412.

21. Ibid., I, pp. 406–7. Emphasis Darwin's.

22. Ibid., I, pp. 408–9. Emphasis Darwin's.

23. Ibid., I, pp. 409–10.

24. Ibid., I, p. 410.

25. Ibid., I, pp. 410–11. Emphasis Darwin's

26. Ibid., I, p. 410.

27. Cf. a June 14, 1856 Darwin letter to Fox. Ibid., I, p. 430.

28. Ibid., I, pp. 442–43. Emphasis Darwin's.

29. Cf. an April 16, 1858 Darwin letter to Fox. Ibid., I, pp. 468–69.

30. Ibid., I, p.413.

31. See Darwin's August 24, 1856 letter to Gray. *More Letters*, I, p. 430.

32. *Life and Letters*, I, p. 420.

33. Ibid., I, pp. 420–24.

34. *More Letters*, I, p. 421.

35. *Life and Letters*, I, p. 437. This view of Darwin's, which I call the least objectionable theory, will be discussed in Chapters 6 and 7. It suggests that Darwin was already willing to see the value of his theory as not so much the correct one, but the least objectionable one.

36. *More Letters*, I, p. 126. The idea that Darwin used other scientists to support his work is not new with me, though I think I am the first to go into any detail about the subject. Cf. Frank N. Egerton, "Darwin's Method or Methods?" "An important aspect of Darwin's methodology . . . is the use which Darwin made of other scientists. He was an amazingly prolific correspondent, and the major portion of this correpondence was with colleagues concerning his research. Professional correspondence is common among scientists, but what seems unique with Darwin is how many scientists he was able to persuade to do research for him. He was a marvelously diplomatic writer, and there appear to have been few scientists who could resist his appeal for assistance. Many of them were so eager to oblige him that they would drop whatever research they were involved in to seek answers to questions. It was almost as if he ran a private research institute, in which some of the ablest scientists were his unsalaried part time assistants."

37. *Life and Letters*, I, p. 121.

38. Ibid., I, pp. 117, 121–22.

39. During the *Beagle* voyage Robert Darwin wrote to Henslow of Charles's letters: "There is a natural good-humoured energy in his letters just like himself." (Ibid., I, p. 191.)

40. Ibid., I, p. 491.

41. *More Letters*, I, p. 425.

42. *Life and Letters*, I, pp. 97–98.

43. Ibid., I, p. 412.

44. Ibid., I, p. 409.

45. Ibid., I, p. 420.

46. "Unpublished Letters," p. 28. He also profusely thanked Robert Fitch,

in his earlier work on Cirripedes. Cf. Trenn, "Charles Darwin, Fossil Cirripedes, and Robert Fitch," pp. 471–90.

Darwin apparently was worried about making too great a show of gratitude. As he wrote in his M Notebook (paragraph 60): "In making too much profession . . . of gratitude . . . I was tending to make myself in *act* less grateful." Emphasis Darwin's. Cf. Gruber and Barrett, *Darwin on Man*, pp. 276, 313.

47. April 10, 1846 letter. *More Letters*, I, p. 414.
48. July 14, 1857 letter. *Life and Letters*, I, p. 461.
49. October 1856 letter to Hooker. Ibid., I, p. 444.
50. February 23, 1858 letter to Hooker. Ibid., I, p. 468.
51. July 8, 1856 letter. *More Letters*, I, p. 426.
52. January 1, 1857 letter. *Life and Letters*, I, p. 447.
53. May 7, 1855 letter. Ibid., I, p. 409.
54. June 8, 1855 letter. Ibid., I, p. 424.
55. Ibid., I, pp. 132–33. I do not think that this was invariably true. For example, toward the end of the *Origin* Darwin gets a bit belligerent toward potential opposition.
56. Ibid., I, pp. 115–16.
57. Ibid., I, pp. 116–17.
58. Ibid., I, pp. 118–19.
59. Ibid., I, p. 119.
60. Ibid., I, p. 413.
61. Ibid., I, p. 410.
62. *More Letters*, I, p. 425.
63. *Life and Letters*, I, p. 125.
64. Francis Darwin, ed., *The Foundations of the Origin of Species: Two Essays Written in 1824 and 1844 by Charles Darwin*. Hereafter referred to as *Foundations*.
65. *Life and Letters*, I, pp. 389–90.
66. Ibid., I, pp. 394–95. Emphasis Darwin's.
67. Ibid., I, p. 406. Emphasis Darwin's.
68. Ibid., I, p. 409. March 27, 1855. Cf. Darwin's April 21, 1856 letter to his naturalist friend Charles Bunbury: "My determination to put difficulties, as far as I can see them, on both sides is a great aid towards candour. . . . " Quoted in R. C. Stauffer, ed. *Charles Darwin's Natural Selection, Being the Second Part of His Big Species Book Written from 1856 to 1858*, p. 529. Hereafter referred to as Long Manuscript.
69. *Life and Letters*, I, p. 384.
70. Ibid., I, pp. 388–89. Emphasis Darwin's.
71. Ibid., I, p. 391.
72. Ibid., I, p. 398. June 13.
73. Ibid., I, p. 404.
74. *More Letters*, I, p. 85.
75. *Life and Letters*, I, p. 395.
76. Ibid., I, p. 437.
77. Ibid., I, p. 439.
78. Ibid., I, p. 441. August 5, 1856 letter. Cf. Darwin's July 8, 1856 letter to

Lyell: " . . . I have been at work, sometimes in triumph, sometimes in despair, for the last month." Quoted in Stauffer, Long Manuscript, p. 9. Several years after publishing the Origin, Darwin wrote to George Bentham, President of the Linnean Society (June 19, 1863): "I remember too well my endless oscillations of doubt and difficulty." (Life and Letters, II, p. 211.)

79. More Letters, I, p. 105.

80. Life and Letters, I, p. 491. Emphasis Darwin's.

81. Ibid., I, pp. 499–500.

82. Ibid., I, p. 503.

83. Ibid., I, p. 505.

84. Ibid., I, p. 526.

85. Ibid, I, p. 527. Emphasis Darwin's.

86. More Letters, I, p. 139. For more on Huxley and evolution, see Erling Eng, "Thomas Henry Huxley's Understanding of 'Evolution.' "

87. Life and Letters, I, pp. 493–94. After Wallace, Darwin was forced to put his ideas together quickly. See Chapters 6 and 8.

88. Ibid., I, p. 409. Other examples with Fox: "Whether I shall do any good I doubt . . . I am getting out of my depth" (May 7, 1855, ibid., I, p. 401); " . . . my notes are so numerous during nineteen years' collection, that it would take me at least a year to go over and classify them" (June 14, 1856, ibid., I, p. 430); "Sometimes I fear I shall break down, for my subject gets bigger and bigger with each month's work" (November 1856, ibid.); "I am like Croesus overwhelmed with my riches in facts" (February 8, 1858, ibid., I, pp. 467–68); "My health has been lately very bad from overwork, and on Tuesday I go for a fortnight's hydropathy . . . my work is everlasting" (April 16, 1858, ibid., I, p. 469).

89. Ibid., I, p. 453.

90. "Unpublished Letters," p. 26.

91. Life and Letters, I, p. 510. For Darwin's state of anxiety over his workload, see Darwin's October 14, 1837 letter to Henslow (Henslow and Darwin, pp. 138–40).

92. Life and Letters,,I, p. 471.

93. Ibid., I, pp. 426–27.

94. Ibid., I, pp. 427–28. Emphasis Darwin's.

95. Ibid., I, p. 486.

96. Ibid., I, p. 319. In some people's eyes he did stand infinitely low following publication of the Origin. See Alvar Ellegård, Darwin and the General Reader: The Reception of Darwin's Theory of Evolution in the British Periodical Press, 1859–1872 for the reception of Darwin's theory among the British popular press; ed. Thomas F. Glick, The Comparative Reception of Darwinism for the reception of Darwin's ideas in various countries, see especially M. J. S. Hodge's article on "England," pp. 3–39; David Hull, Darwin and His Critics: The Reception of Darwin's Theory of Evolution by the Scientific Community for the reception of Darwin's ideas within the scientific community. Much more work will be needed before a definitive understanding of the reception of Darwin's ideas can be reached. Cf. Susan F. Cannon's review of The Comparative Reception of Darwinism.

97. Robert Chambers, Vestiges of the Natural History of Creation. It went

through numerous editions. Cf. Milton Millhauser, *Just Before Darwin, Robert Chambers and Vestiges.*

98. *Life and Letters*, I, p. 399.

99. *More Letters*, I, p. 75.

100. *Life and Letters*, I, p. 494. Emphasis Darwin's.

101. Ibid., I, pp. 494–95.

102. Ibid., I, p. 499.

103. Ibid., I, p. 515.

104. Ibid., I, p. 517. For Owen's views on evolution, see R. M. MacLeod, "Evolutionism and Richard Owen, 1830–1868: An Episode in Darwin's Century."

105. *Life and Letters*, I, p. 495.

106. *More Letters*, II, p. 455.

107. See *Life and Letters*, I, p. 491. 108. *More Letters*, I, p. 119.

108. *More Letters*, I, p. 119.

109. *Life and Letters*, I, p. 522.

110. Ibid., I, p. 513.

111. Ibid., I, p. 522.

112. Ibid., I, p. 526.

113. Ibid.

114. Ibid., I, p. 529. October 23, 1859 letter.

115. Cf. Leonard G. Wilson, *Sir Charles Lyell's Scientific Journals on the Species Question.* Darwin was later greatly disappointed in Lyell's inability to accept his ideas completely. As Darwin wrote to Hooker on February 24, 1863 (*Life and Letters*, II, p. 194): " . . . I must say how much disappointed I am that he has not spoken out on species, still less on man. And the best of the joke is that he thinks he has acted with the courage of a martyr of old." See also *Life and Letters*, II, pp. 196–201, 205–6.

116. *More Letters*, I, p. 119.

117. *Life and Letters*, I, p. 527. Emphasis Darwin's.

118. Ibid., I, p. 530.

119. Ibid., II, p. 20. On the day of publication of the *Origin* (November 24, 1859) Darwin wrote to Huxley: " . . . I wish to know your general impression of the truth of the theory of Natural Selection." See Henry F. Osborn, "A priceless Darwin letter." For an interesting biography of Huxley, see Cyril Bibby, *Thomas Huxley: Scientist, Humanist and Educator.* See also Bibby, "Huxley and the Reception of the 'Origin'."

Michael Bartholomew has reviewed Huxley's defense of the *Origin* in an effort to show that despite his fervor on Darwin's behalf, his advocacy of the case for natural selection was not particularly compelling, and that his own scientific work took no revolutionary new direction after 1859. I think Bartholomew's argument is compelling, if not completely convincing. See Michael Bartholomew, "Huxley's Defense of Darwin." See also T. H. Huxley, *Evolution and Ethics and Other Essays,* and *The Life and Letters of Thomas Henry Huxley,* ed. Leonard Huxley.

120. *Life and Letters*, I, p. 527.

121. Ibid., II, p. 20. Emphasis Darwin's.

Chapter 6

1. We have already seen in Chapter 5 Darwin's doubts about his theory as expressed in his letters. In this chapter, we will see similar concern evident in his pre-*Origin* species writings and notes.

2. Transmutation Notebooks, III, 72.

3. Ibid., II, 202.

4. Ibid., II, 199.

5. Ibid., IV, 160.

6. Ibid., I, 170.

7. Ibid., IV, 47. As Darwin wrote: "Having proved mens & brutes bodies on one type: almost superfluous to consider minds. — as difference between mind of a dog & a porpoise was not thought overwhelming — yet I will not shirk difficulty. . . . "

8. Ibid., IV, 6 — excised pages.

9. Ibid., I, 236.

10. Ibid., I, 217.

11. Ibid., II, 175.

12. Ibid., II, 145.

13. Ibid., II, 176.

14. Ibid., II, 75.

15. Ibid., IV, 5 — excised pages. Emphasis Darwin's. According to de Beer (Transmutation Notebooks, II, p. 29), the problem of the lack of intermediate forms was lessened by Darwin's idea of extinction, yet the idea of extinction was potentially explosive, because one had to assume that Divine Providence had somehow removed its protective blanket from some species.

Many other examples can be cited of Darwin's concern with substantive issues at this time. Compare: (Transmutation Notebooks, II, 222 — excised pages, p. 153) — "Mr. Edw. Blyth . . . thinks passage very rare, in anatomical structure. — the passage between owls & hawks only external . . . intermediate groups often have full structure of one class & full of second — this class if analogous to petrel-grebe external appears to be a puzzle against my theory . . . ;" (ibid., IV, 150–51) — "The weakest part of my theory is the absolute necessity, that every organic being should cross with/another — to escape it in any case we must draw such a monstrous conclusion, that every organ is become fixed & cannot vary — which all facts show to be absurd."

16. Darwin had a good argumentative model in Lyell's very "strategic" organization of the *Principles*. See Alexander M. Ospovat, "The Distortion of Werner in Lyell's *Principles of Geology*" for a discussion of Lyell's well developed strategic capabilities.

Darwin also seemed to have profited strategically from Lyell's attack on Lamarck and Sedgwick's attack on Chambers in terms of preparing his own theory for presentation to the public. See Egerton, "Refutation and Conjecture: Darwin's Response to Sedgwick's Attack on Chambers."

17. See de Beer, Transmutation Notebooks, II, p. 77: Darwin "was . . . concerned with the problem of expressing his views on paper and gave himself in-

structions for presenting his theory." Darwin's concern was such that he was interested in trying to figure out how best to organize his material, and what might be and might not be a good phrase or sentence to use. Concerning organization, Darwin wrote: (ibid., IV, 50–51) " . . . It was absolutely necessary that Physical changes should act not on individuals, but on masses of individuals. — so that the changes should be slow & bear relation to the whole changes of country, & not to the local/changes — this could only be effected by sexes. all the above should follow after discussion of crossing of individuals with respect to representative species, when going North & South"; regarding possible good passages, he wrote: (ibid., III, 22) "A capital passage might be made from comparison of man, with expression of monkey when offended, who loves who fears who is curious . . . who imitates . . . ; (ibid., IV, 23) "A very strong passage might be made — why seeing great variation in external form of varieties, do we suppose bones will not change in *number* [emphasis Darwin's] . . . "; (ibid., III, 36–37) "How beneath the dignity of him, who is supposed to have said let there be light & there was light, — whom it has been declared 'he said let there be light & there was light' — bad taste."

18. Ibid., I, 85. Emphasis Darwin's.

19. Ibid., II, 137.

20. Ibid., IV, 136.

21. "Show independency of shells to external features of *land* . . . " (ibid., II, 99, emphasis Darwin's); "Magazine of Zool. & Bot. Vol. I, p. 456, 4 instances of hybrids between pheasant & Black Fowl, — use as argument possibly some few hybrids in nature . . . " (ibid., II, 184 — excised pages); "Character of analogy, — last acquired, — or aberrant, therefore more easily modified. — This is not easily told. . . . We must argue reversely . . . " (ibid., II, 202); " . . . at the end of *'White's Selbourne'* many references very good. Also *'Rays Wisdom of God.'* Often refer to these . . . " (ibid., II, 248).

Darwin was also concerned about preparing for unfair criticism of his ideas— criticism based on prejudice. He instructed himself to "mention persecution of early Astronomers, — then add chief good of individual scientific men is to push their science a few years in advance only of their age. . . . " Darwin then seemed to try to verbally hold his own feet to the fire: " . . . must remember that if they [i.e., scientific men] *believe* & do not openly avow their belief they do as much to retard as those whose opinion they believe have endeavored to advance cause of truth." (Ibid., II, 123–24 Emphasis Darwin's.)

22. Ibid., IV, 118.

23. Ibid., I, 194.

24. Ibid., III, 21.

25. Ibid., III, 58. Emphasis Darwin's.

26. Ibid., II, 20.

27. Ibid., II, 202. Gavin de Beer has written that to Darwin "the value of a hypothesis increases with the number of facts which it explains" (ibid., I, 104, p. 53). I think the following examples of Darwin's thinking on this issue support de Beer's assessment: (Ibid., III, 117): "The line of argument often pursued throughout my theory is to establish a point as a probability by induction, & to apply it as hypotheses to other points, & see whether it will solve them"; (Ibid.,

III, 71): "In comparing my theory with any other, it should be observed not what comparative difficulties (as long as not overwhelming) [but] what comparative solutions & linking of facts"; (Ibid., I 104): "Absolute knowledge that species die and others replace them. — two hypotheses: fresh creations is mere assumption, it explains nothing further; points gained if any facts are connected." (Ibid., III, 69): " . . . the line of proof & reducing facts to law only merit if merit there be in following work."

Many years later, Darwin wrote to F. W. Hutton (April 20, 1861) revealing his view of proper method and the nature of the evidence he had compiled: "I am actually weary of telling people that I do not pretend to adduce direct evidence of one species changing into another, but that I believe that this view in the main is correct, because so many phenomena can be thus grouped together and explained. . . . I generally throw in their teeth the universally admitted theory of the undulation of light . . . admitted because the view explains so much." See *More Letters*, I, 184.

And a year earlier, at the 1860 British Association debate between Bishop Wilberforce and T. H. Huxley on the value of Darwin's ideas, Huxley used the undulation theory in defense of Darwin's theory. See Chapter 8 for more on this debate.

For more on Darwin and method, see Leon R. Kass, "Teleology and Darwin's *The Origin of Species:* Beyond Chance and Necessity?" in *Organism, Medicine and Metaphysics: Essays in Honour of Hans Jonas*, ed. Stuart F. Spicker; A. C. Crombie, "Darwin's Scientific Method"; James K. Feibleman, "Darwin and Scientific Method."

28. Transmutation Notebooks, II, 267.

29. Ibid., III, 69 and II, 177. Darwin rather modestly admitted that if there was any originality in his ideas, it was a matter of "only slight differences, the opinion of many people in conversation." Darwin felt that "the whole object of the book is its proof," which, of course, made his concern with the strategy of the argument of the *Origin* all that more critical. (See ibid., II, 177.)

On January 18, 1860, soon after publication of the *Origin*, Darwin wrote to Baden Powell that "no educated person, not even the most ignorant, could suppose that I meant to arrogate to myself the origination of the doctrine that species had not been independently created. The only novelty in my work is the attempt to explain *how* species became modified, & to a certain extent how the theory of descent explains certain large classes of facts. . . . " (De Beer, "Some Unpublished Letters," pp. 52–53.) Emphasis Darwin's.

30. Transmutation Notebooks, III, 70. Emphasis Darwin's. For Erasmus's suspected influence on Charles, see Frank N. Egeton, "Darwin's Method or Methods?" pp. 283–84. Michael T. Ghiselin is skeptical of Egerton's claim of Erasmus's influence. See Ghiselin's "Two Darwins: History versus Criticism."

For more on Erasmus Darwin, cf. James Harrison, "Erasmus Darwin's View of Evolution"; Samuel Lilley, "The Origin and Fate of Erasmus Darwin's Theory of Organic Evolution"; D. King-Hele, *The Essential Writings of Erasmus Darwin*, and *Erasmus Darwin*; Nora Barlow, "Erasmus Darwin, F. R. S. (1731–1802)"; N. Garfinkel, "Science and Religion in England, 1790–1800: The Critical Response to the Works of Erasmus Darwin"; Ernst Ludwig Krause, *Eras-*

mus Darwin; Eliza Meteyard, *A Group of Englishmen (1795 to 1815), Being Records of the Younger Wedgwoods and Their Friends.*

31. Transmutation Notebooks, I, 21. I suspect Darwin might also be referring here to his grandfather.

32. Ibid., I, 214.

33. Ibid., I, 216.

34. Ibid., II, 63. For other references to Lamarck, see Ibid., I, 16 and III, 157.

For more on Darwin and Lamarck, see Frank N. Egerton, "Studies of Animal Populations from Lamarck to Darwin," and "Darwin's Early Reading of Lamarck."

35. Transmutation Notebooks, I, 84.

36. Ibid., II, 222 — excised pages.

37. Ibid., II, 199.

38. Ibid., I, 193. As Darwin asked: " . . . would Creator *make* plants when this volcanic point appeared in the great ocean. . . . " Emphasis Darwin's.

39. Ibid., III, 65.

40. Ibid., I, 100. To Darwin also it was "a very great puzzle why Marsupials and Edentata should only have left offsprings in or near South Hemisphere. Were they produced in several places and died off in some? Why did not fossil horse breed in S. America" (Ibid., I, 106). And Darwin noted in mock surprise that " . . . the creative power seems to be checked when islands are near continent. . ." (Ibid., I, 169 — excised pages). Compare also ibid., III, 115.

41. Ibid., I, 218.

42. Ibid., I, 104.

43. Ibid., I, 101.

44. Ibid., I, 216 and III, 36–37. Regarding man, Darwin felt that it was man's arrogance that led him to feel that he was worthy of the attention of a Creator. " . . . man . . . like to think his origin godlike . . .," Darwin wrote (ibid., II, 155). Darwin felt that " . . . man in his arrogance thinks himself a great work worthy the interposition of a deity . . . more humble & I believe truer to/consider him created from animals . . . " (ibid., II, 196–97). " . . .Man acts & is acted on by the organic and inorganic agents of this earth like every other animal," Darwin concluded (ibid., IV, 65).

45. Ibid., I, 115.

46. Ibid., II, 184 — excised pages. Or, as Darwin wrote elsewhere (ibid., IV, 23): "Macleay says it is nonsense to say take a tooth of an animal (as Toxodon) & say its relations, — if we knew its congeners then we can. — Now on my theory this certainly can be accounted for, on any other it is the will of God." See n. 74 below for a description of Macleay's Quinerian theory.

Throughout the Transmutation Notebooks it is evident that Darwin was coming to see some of the positive impact his theory would have. For some of these indications, see ibid., I, 224; II, 135; III, 26; IV, 48 and 71.

47. Ibid., I, 229. Or, as Darwin wrote earlier: " . . . the Creator creates by any laws, which I think is shown by the very facts of the geological character of these islands . . . " (ibid., I, 98).

48. Ibid., II, 166.

49. "Herschel calls the appearance of new species the mystery of mysteries, & has grand passage upon the problem! Hurrah — intermediate causes," Darwin exclaimed with obvious delight. See ibid., IV, 59.

For Darwin on religion, see M. Mandelbaum, "Darwin's Religious Views." See also Robert M. Stecher, "Darwin-Innes Letters: The Correspondence of an Evolutionist with his Vicar, 1848–1884." Most recently, Neal Gillespie has questioned Darwin's vision as thoroughly naturalistic and the man as completely secular. Gillespie feels that the "theological" language in Darwin's writings should not be dismissed as merely poetic or rhetorical. See Gillespie, *Charles Darwin and the Problem of Creation.*

Another important theme in the Transmutation Notebooks which becomes especially prominent in the M and N notebooks is the idea of the continuity among all organisms in nature, including man. Cf. Transmutation Notebooks, I, 73, 232; II, 174; IV, 47.

50. So described by Gruber and Barrett, who also described them as Darwin's "notebooks on man, mind, and materialism." See Gruber and Barrett, *Darwin on Man*, p. xx. Darwin described them as his notebooks on morals and expression.

For Darwin's thinking on such metaphysical issues as the mind/body duality, see C. U. M. Smith, "Charles Darwin, the Origin of Consciousness, and Panpsychism." For Darwin on psychology, see Michael T. Ghiselin, "Darwin and Evolutionary Psychology." For Darwin on man, see Sandra Herbert, "The Place of Man in the Development of Darwin's Theory of Transmutation, Parts I and II."

51. Gruber and Barrett, *Darwin on Man*, M Notebook, 153, p. 296.

52. Ibid., 73, p. 278.

53. Ibid., 125, p. 290. Emphasis Darwin's.

54. Gruber and Barrett, *Darwin on Man*, N Notebook, 65, p. 342. At still another point Darwin advised himself: "If I want some good passages against opposition of divines to progress of knowledge, see Lyell on Scrope." Ibid., 19e, p. 334. See G. P. Scrope, *Memoir on the Geology of Central France*, reviewed by Lyell, *Quarterly Review*. See also Gruber and Barrett, *Darwin on Man*, p. 353.

55. N Notebook, 91, p. 347.

56. M Notebook, 136, p. 292.

57. Ibid., 154e, p. 296. Darwin expands his analysis in the M and N notebooks of the subject of why man finds it necessary to believe in the existence of an intervening Creator. Darwin is impressed with the ideas of the French philosopher Auguste Comte concerning the development of knowledge. Comte felt that each branch of knowledge passes through three successive stages or states of development—the theological or fictitious stage, the metaphysical or abstract stage, and the scientific or positive stage. Darwin felt that the study of natural history at the time fit into the first of Comte's stages. Darwin seems to have learned of Comte's ideas from a review of Comte's *Cours de Philosophie Positive* which appeared in the *Edinburgh Review*. For Darwin's thinking on this subject, cf.: M Notebook, 69, 70, p. 278; 135–36, pp. 291–92; N Notebook, 11–12, p. 332. Darwin felt that John Macculloch in his chapter on the existence of a Deity had an expression "the very same as mine about our origin of a notion of a Deity" (ibid., 35, p. 337). See John Macculloch, *Proofs and Illustrations of the Attributes of God*, Chapter IV, "On the Existence of the Deity. Nature of Proof. Sources of Belief," I, pp. 94–127.

58. N Notebook, 36, p. 337. It was apparently not easy for Darwin to get his contemporaries to believe that the universe could be run without an intervening God. In a letter from Darwin to Lyell (dated August 2, 1861) Darwin complained that Asa Gray and Sir John Herschel clung to the idea of a providential presence in the natural world. "I must think that such views of Asa Gray and Herschel merely show that the subject in their minds is in Comte's theological stage of science," Darwin wrote. (*More Letters*, I, 192) See Gruber and Barrett, *Darwin on Man*, p. 315.

Or compare Darwin's letter to Lyell dated August 21, 1861: "Why should you or I speak of variation as having been ordained and guided, more than does an astronomer, in discussing the fall of a meteoric stone? He would simply say that it was drawn to our earth by the attraction of gravity, having been displaced in its course by the action of some quite unknown laws." (*More Letters*, I, 194.)

59. M Notebook, 72, p. 278.

60. Ibid., 31, p. 271.

61. Ibid., 73, pp. 278–79.

62. Ibid., 57, p. 276. For other examples of Darwin's materialistic thinking, cf. ibid., 19, p. 269; 27, p. 270. See Gruber and Barrett, *Darwin on Man*, pp. 311–12, for more on Darwin's materialistic views.

As in the Transmutation Notebooks, the theme of the great continuity among all organisms is stressed in the M and N notebooks, perhaps with even greater emphasis and attention. See Gruber and Barrett, *Darwin on Man*, pp. 307, 310.

Peter J. Vorzimmer has published for the first time a thirteen-page ink sketch of Darwin's which someone has labelled "1842." Vorzimmer feels that this sketch was "written some time around July of 1839 as a preliminary draft on the subject of 'the principles of variation in animal and vegetable organisms under the effects of domesticity.' " See P. J. Vorzimmer, "An Early Darwin Manuscript: The 'Outline and Draft of 1839.' " More recently, Silvan S. Schweber has suggested that Vorzimmer's "Outline and Draft of 1839" will be identified in a forthcoming article by R. C. Stauffer as the unrevised Chapter I of the 1844 *Essay.* See Schweber, "The Origin of the *Origin* Revisited," p. 230, footnote 5.

63. Darwin, *Foundations.* See also Charles Darwin, *Evolution by Natural Selection,* ed. Gavin de Beer. This latter work includes Francis Darwin's introduction to the 1842 and 1844 *Essays,* the text of both *Essays,* together with the text of the Darwin-Wallace Linnean Society communication of July 1, 1858. All of my references to the *Essays* are to the earlier Francis Darwin edition, with specific reference to either the 1842 or 1844 *Essay.* See note 64, Chapter 5.

64. Darwin, *Foundations,* p. xxii. There is frequently a hedging of statements in the *Essays* with regard to some of Darwin's key concepts. Cf. 1842 *Essay,* p. 23; 1844 *Essay,* pp. 132, 134.

65. See Charles Darwin, *On the Origin of Species by Means of Natural Selection, or the Preservation of Favoured Races in the Struggle for Life,* p. 490. Hereafter referred to as *Origin.* Cf. Morse Peckham's variorum edition: *The Origin of Species: A Variorum Text.* See 1844 *Essay,* p. 52, note 2 for a history of this sentence.

66. 1842 *Essay,* p. 50; 1844 *Essay,* p. 253; *Origin,* p. 485.

67. 1844 *Essay,* p. 253; *Origin,* p. 487.

68. For example, the sentence " . . . the imperfect evidence of the continuousness of the organic series, which, we shall immediately see, is required on
our theory, is against it; and is the most weighty objection . . . " appears in the
1844 *Essay* (p. 134), in a similar form in the 1842 *Essay* (p. 24, note 1), and in
the *Origin* (p. 299); the phrase "and how can we doubt it" in reference to some
of the basic elements of Darwin's theory—struggle, variation, etc.—appears in
the 1844 *Essay* (p. 109) and in a slightly different form in the *Origin* (p. 127):
"and this certainly cannot be disputed."

69. Darwin, *Foundations*, p. xx.

70. Ibid., p. xxv. There are two parts and ten chapters in each *Essay* (but
Francis Darwin divided the 1842 *Essay*, Darwin did not); the 1844 *Essay* expands
sections of the 1842 *Essay* into chapters. The *Origin* is not divided into two
parts. See my discussion of the organization of the *Origin* in Chapter 7.

71. 1842 *Essay*, p. 15.

72. 1844 *Essay*, p. 151. See *Origin*, p. 352.

73. 1844 *Essay*, p. 178.

74. Ibid., p. 202. The Quinerian theory was an ideal approach to classification involving the idea of five major groups of animals, each group having five
subgroups, etc. and all arranged in perfect circles of affinity. See Gruber and
Barrett, *Darwin on Man*, p. 112.

75. 1842 *Essay*, p. 53. Cf. also 1844 *Essay*, p. 145. Emphasis Darwin's.

76. See above.

77. 1844 *Essay*, p. 81. In the *Origin*, p. 45, Darwin insists on variability in
nature, though he seems to understand why some naturalists might not be convinced. As Darwin wrote: "I am convinced that the most experienced naturalist would be surprised at the number of the cases of variability . . . which he
could collect on good authority, as I have collected, during a course of years."

78. 1842 *Essay*, p. 23.

79. 1844 *Essay*, p. 240. Overall, Darwin tends to minimize the occurrence
of variation in nature in the 1842 and 1844 *Essays*. With regard to the former,
Francis Darwin wrote in a footnote: "When the author [Darwin] wrote this sketch
[the *Essay* of 1842] he seems not to have been so fully convinced of the general
occurrence of variation in nature as he afterwards became." (1842 *Essay*, p. 5.)
See also 1844 *Essay*, p. 59, note 1.

80. Ibid., p. 109. In the *Origin* no limit is placed on the extent of variation.

81. Ibid., p. 242. Darwin wrote: " . . . certainly the limit of possible variation of organic beings, either in a wild or domestic state, is not known." (Ibid.,
p. 240.)

82. Ibid., p. 83.

83. Ibid., p. 244.

84. Ibid., p. 134. Similar statements appear in the 1842 *Essay* (p. 24, note 1)
and the *Origin* (p. 299).

85. 1844 *Essay*, pp. 248–49. Darwin wrote: "What evidence is there of a number of intermediate forms having existed, making a passage . . . between the
species of the same groups?" See ibid., p. 136. See *Origin*, p. 301.

86. 1844 *Essay*, p. 243. Darwin continued on the question of the lack of intermediate forms: " . . . we have no reason to suppose that more than a small frac-

tion of the organisms which have lived at any one period have ever been preserved; and hence that we ought not to expect to discover the fossilised sub-
varieties between any two species." (Ibid., p. 244.) A similar point is made in
the *Origin*. See Chapter 7. Or compare Darwin's statement in the 1842 *Essay*,
p. 53: "What arguments against this theory, except our not perceiving every
step, like the erosion of valleys." Compare *Origin*, p. 481: "The difficulty is the
same as that felt by so many geologists, when Lyell first insisted that long lines
of inland cliffs had been formed, and great valleys excavated, by the slow action
of the coast-waves."

87. 1844 *Essay*, p. 254.

88. Ibid.

89. Ibid., p. 145.

90. 1842 *Essay*, p. 31.

91. 1844 *Esssay*, p. 111.

92. Ibid., p. 119.

93. Ibid., p. 81. Emphasis Darwin's. Darwin instructs elsewhere: "Would it
be more striking if we took animals, take Rhinoceros, and study their habitats?
. . . " (1842 *Essay*, p. 30); "N.B. — There ought somewhere to be a discussion
from Lyell to show that external conditions do vary, or a note to Lyell's works
. . . " (ibid., p. 53).

94. Ibid., p. 34.

95. Ibid., p. 7.

96. 1844 *Essay*, p. 81. Emphasis Darwin's.

97. 1842 *Essay*, p. 23.

98. 1844 *Essay*, p. 134. See above.

99. Ibid., p. 112.

100. Ibid., p. 133.

101. Ibid., p. 249.

102. Ibid., p. 133. For more of Darwin on the Creationist view, see 1842 *Essay*,
pp. 33, 50, 52; 1844 *Essay*, pp. 253–55.

103. 1842 *Essay*, p. 51.

104. 1844 *Essay*, p. 251. See a similar passage in the *Origin*, p. 488. For other
examples, see 1842 *Essay*, p. 52; 1844 *Essay*, p. 134. It seems to me that Darwin's point throughout the *Essays* is that there is a structure and unity to the
natural world fitting in with his quest for law idea in the notebooks that cannot be explained by merely saying each individual species has been created.

105. Ibid., p. 248.

106. Ibid., p. 284. Emphasis Darwin's.

107. Ibid., p. 244.

108. Stauffer had difficulties in properly elucidating Darwin's Long Manuscript
text, because of Darwin's drastic revisions which often obscured the continuity of the text and because Darwin's handwriting left many uncertainties. See
Stauffer, Long Manuscript, pp. 2, 18, and 20. See *More Letters*, II, 312, for Darwin's description of his own handwriting: "dreadfully bad." 109. For a discussion of Darwin's argument in the Long Manuscript, see M.J.S. Hodge, "The
Structure and Strategy of Darwin's 'Long Argument'."

110. Long Manuscript, p. 190.

111. Ibid., p. 62. Cf. Ibid., p. 207: "Upon the whole none of the facts, which seem at first to deny that all organic beings have at some period or during some generation to struggle for life are of much weight; on the other hand the several remarks & illustrations given in the foregoing pages, imperfect as they are, appear to me conclusively to show that such struggle, often [of] a very complex nature, does truly exist."

112. Ibid.

113. Ibid., p. 279.

114. Ibid., p. 99. For other examples of Darwin's discussions of ignorance, see ibid., pp. 175, 190, 212.

115. Ibid., p. 205. Darwin wrote: "Another class of facts seemed at one time to me opposed to there being a severe struggle in nature; namely animals having recovered in a state of nature from severe injuries, as evidenced by the fossil Hyaena which had part of its upper jaw entirely worn away. . . . "

116. Ibid., p. 201.

117. Ibid., p. 11.

118. *Life and Letters*, I, p. 443. Cf. also ibid., I, pp. 430, 442.

119. *More Letters*, I, p. 441.

120. Long Manuscript, p. 180. Darwin tried to strike some sort of meaningful balance bewteen completeness and providing an interesting text that would hold the reader's attention. "Have I fairly stated the *more important* objections in *abstract* . . . ," he wrote to Hooker on April 10, 1858 concerning his theory; "to have given all in full would have made my now tedious discussion intolerably tedious." (Quoted ibid., p. 94). Emphasis Darwin's. I am not sure that Darwin was always successful in keeping his text from becoming tedious. Cf. Susan F. Cannon, "Darwin's Vision in *On the Origin of Species*," in *The Art of Victorian Prose*, ed. George Levine and William Madden, p. 155.

121. Long Manuscript, pp. 207, 95.

122. *Life and Letters*, I, p. 494. See Long Manuscript, p. 10.

Chapter 7

1. *Origin*, p. 1.

2. Ibid., p. 2. This fits in with Darwin's overall strategy of candidness, which we saw emphasized in his pre-*Origin* species writings.

3. Ibid.

4. Ibid., p. 4. Because Darwin feels that once people realize how evolution has occurred, they will accept the descent idea.

5. Ibid.

6. Ibid., p. 6.

7. Ibid., p. 12.

8. Ibid., p. 13.

9. Ibid., pp. 30–31.

10. Ibid., p. 10.

11. Ibid. In later editions of the *Origin*, use and disuse became of greater importance. See Peter Vorzimmer, *Charles Darwin: The Years of Controversy, The Origin of Species, and Its Critics, 1859–1882*.

12. *Origin*, pp. 14–15.

13. Ibid., p. 12.

14. Ibid., p. 43.

15. Ibid., p. 18.

16. Ibid., p. 28.

17. Ibid., pp. 47–58.

18. Ibid., p. 47.

19. Ibid., p. 48.

20. Darwin wrote: "I look at individual differences, though of small interest to the systematist, as of high importance for us, as being the first step towards such slight varieties as are barely thought worth recording in works on natural history." (Ibid., p. 51.) Again the common sense problem: People cannot believe what they cannot see.

For more on Darwin and variation, cf. Peter J. Bowler, "Malthus, Darwin, and the Concept of Struggle," and "Darwin's Concepts of Variation." See also Peter J. Vorzimmer, *Charles Darwin: The Years of Controversy.*

21. "I look at the term species, as one arbitrarily given for the sake of convenience to a set of individuals closely resembling each other, and that it does not essentially differ from the term variety. . . . " (*Origin*, p. 52.)

22. Ibid., p. 59.

23. Ibid., p. 44.

24. Ibid., p. 60.

25. Ibid., pp. 60–62. See Thomas Robert Malthus, *An Essay on the Principle of Population, As It Affects the Future Improvement of Society With Remarks on the Speculations of Mr. Godwin, M. Condorcet, and other Writers,* reprinted in *On Population: Thomas Robert Malthus,* ed. Gertrude Himmelfarb.

For more on Darwin and Malthus, see Samuel Levin, "Malthus and the Idea of Progress," and Thomas Cowles, "Malthus, Darwin, Bagehot: A Study in the Transference of a Concept." For Darwin's changing view of adaptation, see Dov Ospovat, "Darwin after Malthus."

26. *Origin*, pp. 62–63. The different between intraspecific and interspecific competition.

27. Ibid., p. 78.

28. " . . . So profound is our ignorance . . . that we marvel when we hear of the extinction of an organic being. . . . " (Ibid., p. 73.)

29. Darwin talks of " . . . our ignorance on the mutual relations of all organic beings. . . . " (Ibid., p. 78.) As we have seen, Darwin uses the term "ignorance" in the sense of people not knowing, not seeing, not having knowledge of.

30. There are several references, as we have seen, to Lyell's works in Darwin's pre-*Origin* species writings and notes. See Chapter 6.

31. Ibid., pp. 95–96.

32. See ibid., pp. 111–15. For an analysis of the concept of extinction in the works of both Darwin and Wallace, see John L. Brooks, "Extinction and the Origin of Organic Diversity."

33. *Origin*, pp. 116–26.

34. Ibid., p. 127. We will remember that in the *Essays* Darwin handled the question of whether or not organisms vary as a question open to some discussion. See Chapter 6.

35. Ibid. In the Long Manuscript, as we have seen, the struggle for existence in nature was handled more as an open question. See Long Manuscript, pp. 172–212.

36. *Origin*, p. 127.

37. Ibid., p. 80.

38. Ibid., p. 96.

39. Ibid., p. 101.

40. Ibid., pp. 101–2.

41. Ibid., p. 107.

42. Ibid., p. 116.

43. Ibid., p. 127.

44. Ibid., p. 97.

45. Ibid., p. 129.

46. Ibid., p. 131.

47. Ibid., p. 132.

48. Ibid., p. 141.

49. Ibid., pp. 143–44.

50. Ibid., p. 166.

51. Ibid., pp. 131–32.

52. Ibid., pp. 162–63.

53. Ibid., p. 152.

54. Ibid., p. 155. One final example: "He who believes that each equine species was independently created, will, I presume, assert that each species has been created with a tendency to vary, both under nature and under domestication . . . ," Darwin writes with regard to the origins of variations. But to admit this view is to Darwin to " . . . reject a real for an unreal, or at least for an unknown, cause. It makes the works of God a mere mockery and deception; I would almost as soon believe with the old and ignorant cosmogonists, that fossil shells had never lived, but had been created in stone so as to mock the shells now living on the sea-shore." (Ibid., p. 167.)

55. Ibid., p. 171.

56. Ibid., pp. 172–79.

57. Ibid., p. 186.

58. Ibid., p. 188.

59. Ibid., p. 195.

60. Ibid., pp. 195–96.

61. Ibid., p. 203.

62. Ibid., p. 204.

63. Ibid., p. 180.

64. Ibid., p. 185.

65. Ibid., p. 199.

66. Ibid., p. 207.

67. Ibid., pp. 227–28.

68. For example, the slave-making instincts of various ants and the fact of neuter or sterile females in insect communities. (See ibid., pp. 224, 236–37.)

69. Ibid., p. 243.

70. Ibid.

71. Ibid., pp. 248–49.

72. Ibid., p. 260. Darwin wants to show how similar species and varieties are, for he wants to get away from the Creationist notion that species are specially created, and varieties the work of secondary laws. Ibid., pp. 275–76.

73. Ibid., p. 272.

74. Ibid., p. 278.

75. Ibid., p. 279.

76. Ibid., p. 280. "Why . . . is not every geological formation and every stratum full of such intermediate links?" Darwin asks. "Geology assuredly does not reveal any such finely graduated organic chain. . . . "

77. Ibid.

78. Ibid.

79. Ibid., p. 301. Darwin also discussed the problems of the sudden appearance of whole groups of allied species in the geological record and the almost entire absence of fossiliferous formations beneath the Silurian strata. See ibid., pp. 302–3, 306–8.

80. Ibid., pp. 310–11. For Darwin and paleontology, see A. S. Romer, "Darwin and the Fossil Record," in *A Century of Darwin*, pp. 130–52, and J. Challinor, "Palaeontology and Evolution," in *Darwin's Biological Work: Some Aspects Reconsidered*, ed. P. R. Bell, pp. 50–100.

81. *Origin*, p. 325.

82. Ibid., p. 327.

83. Ibid., p. 329.

84. Ibid., pp. 346–47. See also ibid., p. 350.

85. Ibid., pp. 352–53.

86. Ibid., p. 355. See Alfred Russel Wallace, "On the Law that has Regulated the Introduction of New Species."

87. See *Origin*, pp. 356–65.

88. Ibid., p. 356.

89. Darwin discusses his Glacial theory in ibid., pp. 365–79.

90. Ibid., pp. 380–81.

91. Ibid., p. 393.

92. Mammals offer Darwin another example. See ibid., p. 389.

93. Ibid., p. 397.

94. See ibid., pp. 398–99.

95. Ibid., p. 408. For more on Darwin and geographical distribution, see P. Darlington, "Darwin and Zoogeography." See also Chapter 4.

96. *Origin*, pp. 414, 420.

97. Ibid., p. 435.

98. Ibid., p. 439.

99. Ibid., p. 443.

100. Ibid., p. 449.

101. Ibid., p. 450. For Darwin and embryology, see Gavin de Beer, "Darwin's Views on the Relations between Embryology and Evolution," and "Darwin and Embryology," the latter in *A Century of Darwin*, pp. 102–29; Jane Oppenheimer, "An Embryological Enigma in the Origin of Species," in *Forerunners of Darwin: 1745–1859*, ed. Bentley Glass et al., pp. 292–322; and Arthur O. Lovejoy, "Recent Criticism of the Darwinian Theory of Recapitulation: Its Grounds and Its Initiator," ibid., pp. 438–58.

102. *Origin*, p. 453.

103. Ibid., pp. 455–56.

104. Ibid., pp. 457–58.

105. We know that he felt that by being forced to write the *Origin* he had a better feeling himself for his theory. See Darwin's October 6, 1858 letter to Hooker quoted above (Chapter 4).

106. Ibid., p. 459.

107. Ibid.

108. Ibid. Included among the difficulties are: lack of transitional forms, lack of fossil remains of transitional forms, the existence of complex and useless organs, instincts, the "universal" sterility of first crosses.

109. Ibid., pp. 465–66.

110. Ibid., pp. 466–67.

111. Ibid., p. 469.

112. Ibid., pp. 469–80.

113. Ibid., p. 480.

114. Ibid., p. 481.

115. Ibid., p. 482.

116. Ibid., pp. 488–89.

117. Ibid., p. 490.

118. This was especially impressive at a time (the second quarter of the nineteenth century) when scientists were demanding more detailed evidence in support of their theories. See Susan F. Cannon, "History in Depth: The Early Victorian Period."

On the question of the evidential basis of Darwin's theory, see Silvan S. Schweber, "The Origin of the 'Origin' Revisited," p. 310.

119. What, for example, would happen if reversion did occur in nature? What then can be learned from the "quite unknown" laws governing inheritance and the possible limits to which variability could go—limits insisted upon by both Malthus and Lyell.

120. See Chapter 5.

121. Darwin on occasion divided his works into two parts: the first to establish the possibility of a theory; the second, to examine direct evidence for or against the theory. See Edward Manier, *The Young Darwin and His Cultural Circle*, pp. 6–7, 151.

122. Cf. Cannon, "Darwin's Vision in *On the Origin of Species*," p. 156.

123. There are, in fact, only a few, fleeting references to Lamarck throughout the *Origin*, even fewer to the *Vestiges'* author; most of the references are of a negative nature. See, for example, the *Origin*, pp. 3–4, 10, 134, 242. For a discussion of the relative value of Lamarck's theory vis-à-vis Darwin's, see C. C. Gillispie, "Lamarck and Darwin in the History of Science."

124. There were other difficulties that Darwin had to contend with in terms of arguing his case. One was the problem that, given a different set of premises, one could use the same facts he had adduced to plead the case for some other theory. We saw Darwin admit this in the "Introduction" to the *Origin*. And we have just seen that there are instances where the Creationist view, if one is willing to accept its premises, seems to provide the "better" explanation. This becomes all the more significant when we realize that Creationist premises were accepted by most naturalists at the time.

An additional problem concerns the various "common sense" questions that could be raised against Darwin's theory. Darwin himself was aware of questions such as the lack of intermediate forms (both living and fossil), the existence of both highly complex and useless organs, the question of how complex organs, such as the eye, could have existed in a previous form, the idea of slow changes adding up to great effects over time, the question of both the existence of variations and the limits to variability, the seeming disparity between the mind of men and animals, the problem of instincts, and the existence of an intense struggle for existence among organisms in nature. Such questions made Darwin's theory as open to attack as the Creationist theory from a "common sense" point of view.

Finally, I think it is significant that the least-objectionable-theory defense of the *Origin* was used by both Huxley and Hooker at the famous British Association Meeting at Oxford University in 1860. *The Athenaeum* reported for July 14, 1860, that, in response to Wilberforce's criticisms, Huxley " . . . without asserting that every part of the theory had been confirmed, . . . maintained that it was the best explanation of the origin of species which had yet been offered . . . " and that Hooker had publicly adopted Darwin's theory "as that which offers by far the most probable explanation of all the phenomena. . . . "

Hooker did, however, add that he was "holding himself ready to lay it down should a better be forthcoming, or should the now abandoned doctrine of original creations regain all it had lost in his experience." See *The Athenaeum*, July 14, 1860, p. 65. For more on the Wilberforce-Huxley debate, which also involved Hooker, see *The Life and Letters of Thomas Henry Huxley*, I, pp. 194–96, 202; and "A Grandmother's Tales," *Macmillan Magazine.*

Chapter 8

1. I suspect Peter Vorzimmer would argue that Darwin never really presented his theory in "finished" form to the public since he seemed to back off so many of his conclusions with succeeding editions of the *Origin*. See Vorzimmer, *Charles Darwin: The Years of Controversy.*

2. *The Autobiography of Charles Darwin*, p. 81. I have always thought that Darwin's fear of standing low in his profession was the other side of his desire to stand at the pinnacle of his profession. He was afraid to fail because he wanted to succeed so much. In this regard, I think Michael Ghiselin has read Darwin's ambition well: " . . . one gets the impression that all his self-abasement conceals a personality hungering for self-esteem." See Ghiselin, *The Triumph of the Darwinian Method*, p. 242.

3. *Life and Letters*, I, p. 341.

4. Ibid., I, p. 352.

5. March 19, 1855. Ibid., I, p. 407. For other indications of Darwin's dissatisfaction with his Cirripedes work, see his June 13, 1849 and October 12, 1849 letters to Hooker (Ibid., I, pp. 397 and 346–47 respectively) and his July 15, 1853 letter to Fox (Ibid., I, p. 335).

6. *The Autobiography of Charles Darwin*, p. 118.

7. "Grand Ideas" in the sense of his 1834 letter to his Cambridge friend Whitely where he talks of finding geology "a never failing interest. . . it cre-

ates the same grand ideas respecting this world which Astronomy does for the universe." See Chapter 2.

8. He won the Copley in 1864. It is interesting that Darwin's *Origin of Species* was noted in the Copley award statement for its "mass of observations," but it was "generally and collectively" expressly omitted from the grounds upon which the award was based. See *Life and Letters*, II, pp. 212–13.

For more on Darwin's receipt of the Copley Medal, see M. S. Bartholomew, "The Award of the Copley Medal to Charles Darwin."

9. The award statement can be found in *Abstracts of the Papers Communicated to the Royal Society of London*, Vol. VI, 1850 to 1854, pp. 355–56.

10. The Royal medal seemed to indicate to Darwin that he had at least attained a modicum of respect from his peers, even though it did not seem to completely satisfy Darwin's hunger for recognition. We get a good idea of what Darwin thought of the award in a letter he wrote to Hooker on November 5, 1854 on the occasion of Hooker's receipt of a Royal medal: " . . . you will find the medal a pleasant little stimulus," Darwin wrote, "when work goes badly, and one ruminates that all is vanity, it is pleasant to have some tangible proof, that others have thought something of one's labours." (*Life and Letters*, I, p. 404.) Darwin received the Wollaston Medal of the Geological Society of London in 1859 for his geological contributions.

11. *Journal of the Linnean Society, Zoology*, III (1858), pp. 45–62. See also Charles Darwin and A. R. Wallace, *Evolution by Natural Selection*. The joint Darwin-Wallace paper apparently had very little impact. See J. W. T. Moody, "The Reading of the Darwin and Wallace papers: An Historical 'Non-event'." See also C. F. A. Pantin, "Alfred Russel Wallace, F. R. S. and His Essays of 1858 and 1855."

12. Cf., for example, Gavin de Beer's description of the incident in *Charles Darwin, Evolution by Natural Selection*, pp. 149–50.

13. Eiseley, *Darwin's Century*, p. 292.

14. *Life and Letters*, I, p. 473.

15. "There is nothing in Wallace's sketch which is not written out much fuller in my sketch, copied out in 1844, and read by Hooker some dozen years ago. About a year ago I sent a short sketch, of which I have a copy, of my views . . . to Asa Gray, so that I could most truly say and prove that I take nothing from Wallace." (Ibid., I, p. 474.)

16. Ibid., I, pp. 474–75.

17. Ibid., I, p. 475.

18. Darwin was often concerned about priority and the chance of losing it, though he seems to have felt guilty about his concern. In a May 3, 1856 letter to Lyell Darwin admitted that "I certainly should be vexed if anyone were to publish my doctrines before me . . . ," though he added: "I rather hate the idea of writing for priority. . ." (Ibid., I, p. 427.) In a May 11, 1856 letter to Hooker, Darwin expressed concern that publishing a preliminary essay might take some novelty away from his intended larger work and " . . . that would grieve me beyond everything." (Ibid., I, p. 429.) After receiving Wallace's sketch, Darwin wrote to Hooker on June 29, 1858: "I dare say all is too late . . . It is miserable in me to care at all about priority." (Ibid., I, p. 476.) In a July 5, 1858 letter to Hooker we get a good indication of Darwin's sensitive feelings on the issue when

he reminds Hooker: " . . . you said you would write to Wallace; I certainly should much like this, as it would quite exonerate me. . ." (Ibid., I, p. 483.) And in a July 13, 1858 letter to Hooker, Darwin wrote: "I always thought it very possible that I might be forestalled, but I fancied that I had a grand enough soul not to care; but I found myself mistaken and punished. . ."(Ibid., I, p. 484.) Finally, in a July 18, 1858 letter to Lyell, Darwin admitted: "I certainly was a little annoyed to lose all priority, but had resigned myself to my fate." (Ibid., I, p. 486.)

19. See Barbara G. Beddall, "Wallace, Darwin and the Theory of Natural Selection," and H. Lewis McKinney, *Wallace and Natural Selection*, Chapter 8, pp. 131–46. As McKinney wrote (p. 143): "[The Wallace affair] . . . is a black mark on the characters of Darwin, Hooker, and Lyell." For a recent analysis of the whole Darwin-Wallace relationship, see Arnold C. Brackman, *A Delicate Arrangement. The Strange Case of Charles Darwin and Alfred Russel Wallace.*

20. Darwin knew that Wallace was working extensively in this area. See H. Lewis McKinney, "Alfred Russel Wallace and the Discovery of Natural Selection." See also McKinney's "Wallace's Earliest Observations on Evolution: 28 December 1845."

21. See Stauffer, Long Manuscript, pp. 1–14.

22. This idea is implied in Barrett and Gruber, *Darwin on Man*, p. 28.

23. Stauffer, Long Manuscript, p. 10. Stauffer also wrote (p. 533): "Page long footnotes did not dismay Victorian publishers, nor discourage thousands of Victorian book buyers. . . . "

24. Cf. *Life and Letters*, I, p. 426.

25. Of course all scientists implicitly employ some sort of management and organizational techniques; it would be very difficult to put any sort of theory together or organize any sort of experiment without some ability to order, organize, and manage. My point with regard to Darwin is that he had these skills developed to an unusual degree; not many scientists have as rigid a living/working schedule as Darwin had at Down, and not many scientists organize a worldwide correspondence network through which, as Frank Egerton has pointed out, other scientists actually perform directed research.

26. Darwin admitted this in the June 26, 1858 letter he wrote to Lyell. As Darwin noted: "Wallace might say, 'You did not intend publishing an abstract of your views till you received my communication. Is it fair to take advantage of my having freely, though unasked, communicated to you my ideas, and thus prevent me forestalling you?' " (Ibid., I, p. 475.)

27. Darwin wrote to Hooker on July 13, 1858: "I am *more* than satisfied at what took place at the Linnean Society. I had thought that your letter and mine to Asa Gray were to be only an appendix to Wallace's paper." (Ibid., I, p. 484.) Emphasis Darwin's.

28. *J. Hist. Biol.* 8 (1975):243–73. Other articles that should be consulted regarding the bibliography of Darwin studies include: Bert James Loewenberg, "Darwin and Darwin Studies, 1959–1963," and Michael Ruse, "The Darwin Industry—A Critical Evaluation." Loewenberg provides an excellent survey of the literature generated by the Darwin centenary while Ruse is critical of recent Darwin scholarship for being too narrowly focused. See also Ernst Mayr, "Open

Problems of Darwin Research." Ruse has recently published a work that attempts to provide a broad overview of the Darwinian period. See Michael Ruse, *The Dawinian Revolution*.

29. There is a fourth, smaller category that Greene describes—the Darwin buffs, the skilled amateurs.

30. *The Great Chain of Being: A Study of the History of an Idea*; also Lovejoy, "Some Eighteenth Century Evolutionists."

31. Greene, "Reflections on the Progress of Darwin Studies," pp. 250–51.

32. Cf. Barzun, *Darwin, Marx and Wagner: Critique of a Heritage*; Loewenberg, "The Mosaic of Darwinian Thought," pp. 3–18; Gillispie, *The Edge of Objectivity: An Essay in the History of Scientific Ideas*; Dupree, *Asa Gray, 1810–1888*; Himmelfarb, *Darwin and the Dawinian Revolution*; Fleming, "Charles Darwin, the Anaesthetic Man"; Young, "Malthus and the Evolutionists: The Common Context of Biological and Social Theory"; and Greene, *The Death of Adam*, and "Darwin as a Social Evolutionist."

33. Greene, "Reflections on the Progress of Darwin Studies," p. 258.

34. See Ghiselin, *The Triumph of the Darwinian Method*.

35. See Mayr, "Lamarck Revisited"; M.J.S. Rudwick, *The Meaning of Fossils: Episodes in the History of Paleontology*; and Gruber, *Darwin on Man*. See also Mayr's "Darwin and Natural Selection."

36. Cf. Barnett, ed., *A Century of Darwin*; Bell, ed., *Darwin's Biological Work: Some Aspects Reconsidered*; Darlington, *Darwin's Place in History*; de Beer, *Charles Darwin, Evolution by Natural Selection*; L. Eiseley, *Darwin's Century*; and Simpson, *The Meaning of Evolution*.

37. Cf. Canguilhem, *Études d'histoire et de philosophie des sciences*; Limoges, *La selection naturelle*. Conry wrote a dissertation on the reception of Darwin's writings in France, while Cadieux wrote one on the intellectual relationship among A. L. de Jussieu, Lamarck, and Cuvier.

38. Cf. Young, "Malthus and the Evolutionists: The Common Context of Biological and Social Theory"; Rudwick, *The Meaning of Fossils*; Hodge, "The Universal Gestation of Nature: Chambers' *Vestiges* and *Explanations*"; Bynum, "The Anatomical Method, Natural Theology, and the Functions of the Brain"; Bartholomew, "Lyell and Evolution: An Account of Lyell's Response to the Prospect of an Evolutionary Ancestry for Man."

39. Cf. Burckhardt, "The Inspiration of Lamarck's Belief in Evolution"; Farber, "Buffon and the Concept of Species"; McKinney, *Wallace and Natural Selection*; P. Vorzimmer, *Charles Darwin, the Years of Controversy*, and "Charles Darwin and Blending Inheritance"; Stauffer, Long Manuscript; Cannon *Science in Culture: The Early Victorian Period*, and "The Normative Role of Science in Early Victorian Thought"; Coleman, *Biology in the Nineteenth Century: Problems of Form, Function, and Transformation*; Oppenheimer, *Essays in the History of Embryology and Biology*; Stocking, "From Chronology to Ethnology: James Cowles Prichard and British Anthropology 1800–1850," in James C. Prichard, *Researches into the Physical History of Man*.

For an analysis of some of the non-scientific origins of Victorian evolutionary ideas, see John W. Burrow, *Evolution and Society: A Study in Victorian Social Theory*.

40. Barzun, *Darwin, Marx, and Wagner*, p. 77.
41. Ibid., p. 80.
42. Ibid.
43. Ibid.
44. Ibid., pp. 80–81.
45. Ibid., p. 81.
46. Ibid.
47. Ibid., p. 82.
48. Ibid.
49. Ibid.
50. Ibid., p. 92.
51. Ibid., p. 94. Barzun also criticizes Darwin for disregarding his predecessors, suggesting that in Darwin's "Historical Sketch" (included in the second and following editions of the *Origin*) Darwin dismissed Buffon, misrepresented Lamarck, and buried Erasmus Darwin in a footnote. Ibid., pp. 88–89.
52. Darlington, *Darwin's Place in History*, pp. 26, 31, 59. Cf. P. G. Mudford, "Lawrence's Natural History of Man (1819)."
53. Darlington, *Darwin's Place in History*, pp. 22–26.
54. Ibid., p. 27.
55. Ibid., p. 63.
56. Ibid., p. 60. Cf. also Darlington, "The Origin of Darwinism." For another disparaging description of Darwin's talents, see William Irvine, *Apes, Angels, Victorians*, pp. 70–71.
57. Himmelfarb, *Darwin and the Darwinian Revolution*, p. 436.
58. Ibid., pp. viii and 333.
59. Ibid., pp. 334–35. William Whewell criticized Darwin on the possibility of assuming intermediate steps: "For it is assumed that the mere possibility of imagining a series of steps of transition from one condition of organs to another, is to be expected as a reason for believing that such transition has taken place. And next, that such a possibility being thus imagined, we may assume an unlimited number of generations for the transition to take place in, and that this indefinite time may extinguish all doubt that the transitions really have taken place." *Astronomy and General Physics*, pp. xvii–xviii. Quoted in Himmelfarb, *Darwin and the Darwinian Revolution*, pp. 333–34. Cf. also Cannon, "Darwin's Vision in *On the Origin of Species*," pp. 169–72 for a critique of Darwin's scientific logic.
60. Himmelfarb, *Darwin and the Darwinian Revolution*, p. 446.
61. Eiseley, *Darwin's Century*, pp. 148–56.
62. Ibid., pp. 155, 335–36. As Eiseley writes (ibid., p. 336): " . . . neither Darwin nor his immediate followers seem to have had any particular feeling for the internal stability and harmony of the organism. Their success with the concept of struggle in the exterior environment had led them to see everything through this set of spectacles."
63. Ibid., p. 216.
64. Gillispie, *The Edge of Objectivity*, p. 304.
65. Ibid., p. 342.
66. Ibid., p. 338.

67. Ibid., p. 342.

68. Ibid., p. 305.

69. Ghiselin, *The Triumph of the Darwinian Method*, p. 2.

70. Ibid., p. 4.

71. Ibid., p. 5. Ghiselin writes: Darwin's "philosophical conception of the scientific method was quite sophisticated."

72. Ibid., p. 3.

73. There is considerable argument among philosophers of science concerning whether in fact Darwin's arguments regarding natural selection fit the logical empiricist hypothetico-deductive model. For those who say they do, see A.G.N. Flew, "The Structure of Darwinism"; A. C. Crombie, "Darwin's Scientific Method"; H. Lehman, "On the Form of Explanation in Evolutionary Theory."

For those who say they do not, see A. R. Manser, "The Concept of Evolution," and A. D. Barker, "An Approach to the Theory of Natural Selection."

Michael Ruse takes a middle ground position. He feels that although one might argue that in a sketchy sort of way a large part of Darwin's theory is hypothetico-deductive in form, the "structure is . . . intermeshed with a great many analogical threads." Ruse feels that it " . . . is undeniable that, as given, Darwin's arguments are not rigorously deductively valid." I find Ruse's analysis convincing. See Ruse, "Charles Darwin's Theory of Evolution: An Analysis." See also Ruse, "Natural Selection in the *Origin of Species.*"

74. Cf. Huxley, *Evolution, The Modern Synthesis*; Simpson, *The Meaning of Evolution*; G. de Beer, *Charles Darwin, Evolution by Natural Selection*.

75. See Cannon, "The Bases of Darwin's Achievement: A Reevaluation," pp. 111, 133; "Charles Lyell, Radical Actualism, and Theory," p. 119; "The Whewell-Darwin Controversy," p. 382; and *Science in Culture*, p. 87.

76. Cannon, "Uniformitarian-Catastrophist Debate," pp. 52–53. See also Cannon, "The Bases of Darwin's Achievement," p. 133.

77. See Cannon, "Uniformitarian-Catastrophist Debate," p. 55, and "The Bases of Darwin's Achievement," p. 117. To Cannon, Lyell was "the scientist he [Darwin] had to beat." ("Charles Lyell, Radical Actualism, and Theory," p. 119); Darwin's "theory is specifically designed to refute Lyell" ("The Whewell-Darwin Controversy," p. 378).

78. Ibid., pp. 377, 380–83. For the idea that Darwin's theory more than just supplanted the idea of beneficient design but also supplanted the whole idealist movement in biology, see Peter J. Bowler, "Darwinism and the Argument from Design: Suggestions for a Reevaluation."

79. Cannon, "The Whewell-Darwin Controversy," p. 383.

80. Cannon's introduction in Vorzimmer, *Charles Darwin, the Years of Controversy*, p. xiv. Cannon wrote in "Darwin's Vision in *On the Origin of Species*," p. 155: "Darwin used all the devices readily available to him to make the argument effective."

81. Cannon's introduction in Vorzimmer, *Charles Darwin, the Years of Controversy*," p. xv. She wrote in "Darwin's Vision in *On the Origin of Species*," p. 171: " . . . it was the intuitions of the *Origin*—the intuitions, not the evidence, for there is none; and not the logic, for it is poor—it was Darwin's intuitions as

expressed in the *Origin* which transformed biology and geology." See also Cannon, "The Bases of Darwin's Achievement," p. 134, and *Science in Culture*, p. 280. For a review of some of Cannon's recent ideas, see Silvan S. Schweber, "Early Victorian Science: *Science in Culture.*"

82. Gruber and Barrett, *Darwin on Man.*

83. Vorzimmer, *Charles Darwin, the Years of Controversy*, p. xviii.

84. Limoges, *La selection naturelle.* I would also include within this neutral category a group of scholars whose work has appeared primarily in the pages of the *Journal of the History of Biology* within recent years. These scholars, basing their work on archival research on the Darwin papers deposited at the University of Cambridge Library, have sought to reconstruct in detail the changing nature of Darwin's theory in the years between Darwin's return from the *Beagle* voyage and the publication of the *Origin* as well as the precise processes by which that theory developed during those years. I cannot say at this point that one particular view of Darwin and his work has emerged from these valuable researches, though this work may well provide the basis for a new interpretation of Darwin in the future. Within this archival group I would include such Darwin scholars as Janet Browne, Sandra Herbert, Dov Ospovat, R. Alan Richardson, Silvan Schweber, and Frank Sulloway.

85. Personally, I have always had the most trouble following Darwin's discussion of divergence of character in Chapter IV of the *Origin.* For some thoughts on the confusing nature of parts of the argument of the *Origin*, see Hodge, "The Structure and Strategy of Darwin's 'Long Argument,' " pp. 242–45. For more on Darwin's theory of divergence, see Janet Browne, "Darwin's Botanical Arithmetic and the 'Principle of Divergence,' 1854–1858," and Silvan S. Schweber, "Darwin and the Political Economists: Divergence of Character."

86. Some examples of what I consider weak arguments in the *Origin*: (1) In Chapter VI of the *Origin*, where, as we have seen, Darwin attempts to deal with the problem of natural selection having formed organs as complex as the eye, he writes that although the idea that an organ so perfect as the eye was formed by natural selection "is more than enough to stagger anyone," there is yet "no logical impossibility in the argument of [achieving] any conceivable degree of perfection through natural selection." Although Darwin is certainly correct in suggesting that there is "no logical impossibility" to the degree of perfection an organ might achieve under the natural selection argument, the real question here is not the argument's lack of logical *impossibility*, but whether the occurrence of certain organs or characteristics is very *probable* under it. (2) We have seen Darwin's use of the absolutely fatal objection idea. The problem that I see with Darwin's tendency to view objections as being fatal to his theory is that once he has shown that they are not absolutely fatal—of course, few objections really would be—there is sometimes a tendency not to assess their less than "fatal" impact on his theory. For example, though an objection might not be fatal to his theory, it might nonetheless shed considerable doubt on its validity. I do not think that we should be expected to equate "not being fatal" with "no longer being a problem," as sometimes seems to be Darwin's implication. (3) Darwin suggests in Chapter VI of the *Origin* that one proof that transitional forms have existed can be found in the age-old doctrine of "*natura non facit*

saltum," nature makes no leaps. We meet with this admission, Darwin argues, in almost all experienced contemporary naturalist writings (*Origin*, pp. 194, 206). Darwin's clear implication is that as a result we should tend to accept the validity of the doctrine. But it is also true, as Darwin mentions in his introduction to the *Origin* and admits again in Chapter IX, that in almost all experienced contemporary naturalist writings the doctrine of the immutability of species is upheld. Are we to believe these experienced naturalists when it comes to the doctrine of *"natura non facit saltum"* and disbelieve them on the doctrine of the immutability of species?

87. For an account of the development of Darwin's theory of Pangenesis, see G. L. Geison, "Darwin and Heredity: The Evolution of his Hypothesis of Pangenesis."

88. Greene, "Reflections on the Progress of Darwin Studies," p. 254.

Works Cited

Primary Sources

Barrett, Paul H. "A Transcription of Darwin's First Notebook on 'Transmutation of Species.' " *Bulletin of the Harvard Museum of Comparative Zoology* 122 (1960):245–96.

"The British Association." *The Athenaeum*, July 14, 1860, p. 65.

Chambers, Robert. *Vestiges of the Natural History of Creation*. London: J. Churchill, 1844.

Comte, Auguste. *Cours de Philosophie Positive*. 8 volumes. Paris: 1830–35.

Darwin, Charles. "Geological Notes Made During a Survey of the East and West Coasts of South America, in the Years 1832, 1833, 1834, and 1835, With an Account of a Transverse Section of the Cordillera of the Andes between Valparaiso and Mendoza." Communicated by Professor Adam Sedgwick. *Proceedings of the Geological Society of London* II (1833–38):210–12.

———. "Observations of Proofs of Recent Elevation on the Coast of Chile." *Proceedings of the Geological Society of London* II (1833–38):446–49.

———. "On Certain Areas of Elevation and Subsidence in the Pacific and Indian Oceans as Deduced from the Study of Coral Formations." *Proceedings of the Geological Society of London* II (1833–38):552–54.

———. "On the Connexion of Certain Volcanic Phenomena." *Proceedings of the Geological Society of London* II (1833–38): 654–60.

———. *Notebooks on Transmutation of Species*
Part I: *First Notebook* (July 1837–February 1838), edited with an introduction and notes by Gavin de Beer. *Bulletin of the British Museum (Natural History)*, Historical Series, 1960, vol. 2, no. 2.
Part II: *Second Notebook* (February 1838–July 1838). Ibid., 1960, vol. 2, no. 3.
Part III: *Third Notebook* (July 15, 1838–October 2, 1838). Ibid., 1960, vol. 2, no. 4.

Part IV: *Fourth Notebook* (October 1838–July 10, 1839). Ibid., 1960, vol. 2, no. 5.

Part V: *Addenda and Corrigenda,* edited with notes by Gavin de Beer and M. J. Rowlands. Ibid., 1961, vol. 2, no. 6.

Part VI: *Pages Excised by Darwin,* edited by Gavin de Beer, M. J. Rowlands, and B. M. Skramovsky. Ibid., 1967, vol. 3, no. 5.

————. "Observations on the Parallel Roads of Rob Roy, and other Parts of Lochaber in Scotland, with an attempt to prove that they are of marine origin." *Philosophical Transactions of the Royal Society of London* (1839):39–81.

————. *Journal of Researches into the Geology and Natural History of the Various Countries Visited by H.M.S. Beagle.* London: Henry Colburn, 1839.

————. *Zoology of the Voyage of H.M.S. Beagle.* Edited and superintended by Charles Darwin. 5 parts. London: Smith, Elder & Co., 1839–43.

Part I: *Fossil Mammalia,* by Richard Owen, with a geological introduction by Charles Darwin, 1840.

Part II: *Mammalia,* by George R. Waterhouse, with a notice of their habits and ranges by Charles Darwin, 1839.

Part III: *Birds,* by John Gould, 1841.

Part IV: *Fish,* by Rev. Leonard Jenyns, 1842.

Part V: *Reptiles,* by Thomas Bell, 1843.

————. *Questions About the Breeding of Animals [1840].* Edited with an introduction by Gavin de Beer. London: Society for the Bibliography of Natural History, 1968.

————. *The Foundation of the Origin of Species: Two Essays Written in 1842 and 1844 by Charles Darwin.* Edited by Francis Darwin. Cambridge: University Press, 1909.

————. *Geological Observations on Volcanic Islands and Parts of South America.* 3rd ed. New York: D. Appleton, 1900.

————. *A Monograph of the Fossil Lepadidae, or Pedunculated Cirripedes of Great Britain.* London: Palaeontographical Society, 1851.

————. *A Monograph of the Sub-Class Cirripedia, with figures of all Species. The Lepadidae, or Pedunculated Cirripedes.* London: Ray Society, 1851.

————. *A Monograph of the Sub-Class Cirripedia. The Balanidae . . . The Verruciade.* London: Royal Society, 1854.

————. *A Monograph of the Fossil Balanidae and Verruciade of Great Britain.* London: Palaeontographical Society, 1854.

————. *Charles Darwin's Natural Selection, Being the Second Part of His Big Species Book Written from 1856 to 1858.* Edited from the manuscript by R. C. Stauffer. Cambridge: Cambridge University Press, 1975.

————. *On the Origin of Species by Means of Natural Selection, or the Preservation of Favoured Races in the Struggle for Life.* London: John Murray, 1859. Facsimile edition. New York: Athenaeum, 1967.

————. *The Origin of Species: A Variorum Text.* Prepared by Morse Peckham. Philadelphia: University of Pennsylvania Press, 1959.

―――. *The Autobiography of Charles Darwin, 1809–1882, With Original Omissions Restored.* Edited by Nora Barlow. New York: Harcourt Brace, 1958.

―――. *The Life and Letters of Charles Darwin, Including an Autobiographical Chapter; Edited by his Son Francis Darwin.* 2 vols. New York: D. Appleton, 1894.

―――. *More Letters of Charles Darwin: A Record of His Work in a Series of Hitherto Unpublished Letters.* 2 vols. Edited by Francis Darwin and A. C. Seward. New York: D. Appleton, 1903.

―――. "Some Unpublished Letters of Charles Darwin." Edited by Gavin de Beer. *Notes and Records of the Royal Society of London* 14 (1959):12–66.

―――. "Further Unpublished Letters of Charles Darwin." Edited by Gavin de Beer. *Annals of Science* 14 (1958):83–111.

―――. *Darwin and Henslow: The Growth of An Idea.* Edited by Nora Barlow. London: John Murray, 1967.

―――. "The Darwin Letters at Shrewsbury School." Edited by Gavin de Beer. *Notes and Records of the Royal Society of London* 23 (1968):68–85.

―――. "Darwin Journal." Edited by Gavin de Beer. *Bulletin of the British Museum (Natural History)*, Historical Series 2 (1959):1–21.

Darwin, Charles and Wallace, A. R. *The Darwin-Wallace Celebration Held on Thursday, 1st July, 1908 by the Linnean Society of London.* London: Linnean Society, 1908.

―――. *Evolution by Natural Selection.* Foreword by Gavin de Beer. Cambridge: Cambridge University Press, 1958. Includes Francis Darwin's introduction to the 1842 and 1844 *Essays* and the text of the *Essays* together with the text of the Darwin-Wallace Linnean Society communication of July 1, 1858.

"Darwin-Wallace Joint Communication to the Linnean Society, July 1, 1858." *Journal of the Linnean Society, Zoology* III (1858):45–62.

Darwin, Emma. *Emma Darwin: A Century of Family Letters, 1792–1896.* 2 vols. Edited by Henrietta Litchfield. New York: D. Appleton, 1915.

Darwin, Leonard. "Memories of Down House." *The Nineteenth Century and After* 106 (1929):118–23.

Forbes, Edward. "On the Connexion between the Distribution of the Existing Fauna and Flora of the British Isles and the Geological Changes which have affected their area." *Memoirs of the Geological Survey of Great Britain* 1 (1846):336–432.

"A Grandmother's Tales." *Macmillan Magazine* LXXVIII (1898):433.

Herschel, John F. W. *Preliminary Discourse on the Study of Natural Philosophy.* London: Longman, Rees, et al., 1831.

Hooker, Joseph Dalton. *The Botany of the Antarctic Voyage of H.M. Discovery Ships Erebus and Terror in the Years 1839–1843, Under the Command of Captain Sir James Clark Ross.* 6 vols. London: Reeve Bros., 1844–60.

―――. *Himalayan Journals: or, Notes of a Naturalist in Bengal, the*

Sikkim and Nepal Himalayas, the Khasia Mountains, &c. 2 vols. London: John Murray, 1854.

——. *The Life and Letters of Sir Joseph Dalton Hooker.* Edited by Leonard Huxley. 2 vols. London: John Murray, 1918.

Hooker, Joseph Dalton, and Thomson, Thomas. *Introductory Essay to the Flora Indica.* London: W. Pamplin, 1855.

Huxley, T. H. *Evolution and Ethics and Other Essays.* New York: D. Appleton, 1899.

——. *The Life and Letters of Thomas Henry Huxley.* Edited by Leonard Huxley. 2 vols. New York: D. Appleton, 1901.

King-Hele, D. *The Essential Writings of Erasmus Darwin.* London: Mac-Gibbon & Kee, 1968.

Lyell, Charles. Review of G. P. Scrope's *Memoir on the Geology of Central France. Quarterly Review* 36 (1827):437–83.

——. *The Principles of Geology, Being an Attempt to Explain the Former Changes of the Earth's Surface by Reference to Causes Now in Operation.* 3 vols. London: John Murray, 1830–33.

——. "Presidential Address to the Geological Society, 1836." *Proceedings of the Geological Society of London* II (1833–38):367.

——. "Presidential Address to the Geological Society, 1837." *Proceedings of the Geological Society of London* II (1833–38):504–6.

——. *The Life, Letters and Journals of Charles Lyell.* Edited by his sister–in–law, Mrs. Lyell. 2 vols. London: John Murray, 1881.

Macculloch, John. *Proofs and Illustrations of the Attributes of God.* 3 vols. London: Duncan, 1837.

Malthus, Thomas Robert. *An Essay on the Principle of Population, As it Affects the Future Improvement of Society, with Remarks on the Speculations of Mr. Goodwin, M. Condorcet and Other Writers.* Published anonymously 1798; reprinted in *On Population: Thomas Robert Malthus.* Gertrude Himmelfarb, ed. New York: Modern Library, 1960.

Macgillivray, William. *A History of British Birds, Indigenous and Migratory, Including Their Organization, Habits, and Relations.* 5 vols. London: Scott, Webster, Geary, 1837–52.

Meteyard, Eliza. *A Group of Englishmen (1795–1815) Being Records of the Younger Wedgwoods and Their Friends.* London: Longmans, Green, 1871.

Osburn, Henry Fairfield. "A Priceless Darwin Letter." *Science* LXIV (1926):476–77.

Pritchard, James C. *Researches into the Physical History of Man.* Edited with an introduction by George W. Stocking, Jr. Chicago: University of Chicago Press, 1973.

Rosse, Earl of. "Copley Award Statement. 1854." *Abstracts of the Papers Communicated to the Royal Society of London* VI (1850–54):355–56.

Scrope, G. P. *Memoir on the Geology of Central France.* London: John Murray, 1827.

Sedgwick, Adam. "Address to the Geological Society, Delivered on the Evening of the Anniversary, February 18, 1831." *Proceedings of the Geological Society of London* I (1831):281–316.

Trenn, Thaddeus J. "Charles Darwin, Fossil Cirripedes, and Robert Fitch: Presenting Sixteen Hitherto Unpublished Darwin Letters of 1849 to 1851." *Proceedings of the American Philosophical Society* 118 (1974):471–89.

Vorzimmer, Peter J. "An Early Darwin Manuscript: The 'Outline and Draft of 1839.' " *Journal of the History of Biology* 8 (1975):191–217.

Wallace, Alfred. "On the Law that has Regulated the Introduction of New Species." *Annals and Magazine of Natural History* 16 (1855):184–96.

Whewell, William. Review of Lyell's *Principles. The British Quarterly Theological Review* 9 (1831):180–206.

———. *Astronomy and General Physics.* London: W. Pickering, 1864.

Secondary Sources

Adler, S. "Darwin's Illness." *Nature* 184 (1959):1102–3.

Allan, Mea. *The Hookers of Kew, 1785–1911.* London: Joseph, 1967.

———. *Darwin and His Flowers: The Key to Natural Selection.* New York: Taplinger, 1977.

Ashworth, J. H. "Charles Darwin as a Student in Edinburgh." *Proceedings of the Royal Society of Edinburgh* 65 (1935):97–113.

Atkins, Hedley. *Down, the Home of the Darwins: The Story of a House and The People Who Lived There.* London: Royal College of Surgeons of England, 1974.

Bailey, Sir Edward. *Charles Lyell.* Garden City, New York: Doubleday, 1963.

Barker, A. D. "An Approach to the Theory of Natural Selection." *Philosophy* 44 (1969):271–90.

Barlow, Nora. "Robert Fitzroy and Charles Darwin." *Cornhill Magazine* 72 (1932):493–510.

———. "Erasmus Darwin, F. R. S. (1731–1802)." *Notes and Records of the Royal Society of London* 14 (1959):85–98.

Barnett, S. A. ed. *A Century of Darwin.* Cambridge, Mass.: Harvard University Press, 1958.

Barrett, Paul H. "The Sedwick-Darwin Geologic Tour of North Wales." *Proceedings of the American Philosophical Society* 118 (1974):146–64.

Bartholomew, M. J. "The Award of the Copley Medal to Charles Darwin." *Notes and Records of the Royal Scoiety of London* 30 (1976):209–18.

Bartholomew, Michael. "Lyell and Evolution: An Account of Lyell's Response to the Prospect of an Evolutionary Ancestry for Man." *British Journal for the History of Science* 6 (1973):261–303.

———. "Huxley's Defense of Darwin." *Annals of Science* 32 (1975):525–35.

———. "The Non-progress of Non-progression: Two Responses to Lyell's Doctrine." *British Journal for the History of Science* 9 (1976):166–74.

Barzun, Jacques. *Darwin, Marx, and Wagner: Critique of a Heritage.* Boston: Little, Brown, 1941.

Basalla, George. "The Voyage of the *Beagle* without Darwin." *Mariner's Mirror* 49 (1963):42–48.

218   *Bibliography*

Beddall, Barbara G. "Wallace, Darwin, and the Theory of Natural Selection." *Journal of the History of Biology* 1 (1968):261–324.

———. "Wallace, Darwin, and Edward Blyth: Further Notes on the Development of Evolution Theory." *Journal of the History of Biology* 5 (1972):153–58.

———. "'Notes for Mr. Darwin': Letters to Charles Darwin from Edward Blyth at Calcutta: A Study in the Process of Discovery." *Journal of the History of Biology* 6 (1973):69–95.

de Beer, Gavin. "Darwin's Views on the Relations between Embryology and Evolution." *Journal of the Linnean Society of London, Zoology* 44 (1958):12–23.

———. "The Origins of Darwin's Ideas on Evolution and Natural Selection." *Proceedings of the Royal Society of London* 155 (1962):321–28.

———. *Charles Darwin: Evolution by Natural Selection.* London: Thomas Nelson, 1963.

Bell, P. R., ed. *Darwin's Biological Work: Some Aspects Reconsidered.* Cambridge:Cambridge University Press, 1959.

Bibby, Cyril. *Thomas Huxley: Scientist, Humanist, and Educator.* London: Watts, 1959.

———. "Huxley and the Reception of the 'Origin.' " *Victorian Studies* 3 (1959):76–86.

Bowler, Peter J. "Darwin's Concepts of Variation." *Journal of the History of Medicine* 29 (1974):196–212.

———. *Fossils and Progress: Palaeontology and the Idea of Progressive Evolution in the Nineteenth Century.* New York: Science History Publications, 1976.

———. "Malthus, Darwin and the Concept of Struggle." *Journal of the History of Ideas* 37 (1976):631–50.

———. "Darwinism and the Argument from Design: Suggestions for a Reevaluation." *Journal of the History of Biology* 10 (1977):29–43.

Brackman, Arnold C. *A Delicate Arrangement. The Strange Case of Charles Darwin and Alfred Russel Wallace.* New York: Times Books, 1980.

Brooks, John L. "Extinction and the Origin of Organic Diversity." *Transactions of the Connecticut Academy of Arts and Sciences* 44 (1972):19–56.

Browne, Janet. "The Charles Darwin-Joseph Hooker Correspondence: An Analysis of Manuscript sources and Their Use in Biography." *Journal of the Society for the Bibliography of Natural History* 8 (1978):351–66.

———. "Darwin's Botanical Arithmetic and the 'Principle of Divergence,' 1854–1858." *Journal of the History of Biology* 13 (1980):53–89.

Brussel, James A. "The Nature of the Naturalist's Unnatural Illness: A Study of Charles Robert Darwin." *Psychiatric Quarterly Supplement*, pt. 2, 1966, pp. 1–7.

Burchfield, Joe D. "Darwin and the Dilemma of Geologic Time." *Isis* 65 (1974):301–21.

Burckhardt, Richard. "The Inspiration of Lamarck's Belief in Evolution." *Journal of the History of Biology* 5 (1972):413–38.

Burrow, J. W. *Evolution and Society: A Study in Victorian Social Theory.* Cambridge: Cambridge University Press, 1966.

Burstyn, Harold. "If Darwin Wasn't the *Beagle*'s Naturalist, Why was He on Board?" *British Journal for the History of Science* 8 (1975):62–69.

Bynum, William. "The Anatomical Method, Natural Theology, and the Functions of the Brain." *Isis* 64 (1973):445–68.

Canguilhem, Georges. *Études d'histoire et de philosophie des sciences.* Paris: Vrin, 1970.

Cannon, Susan F. "The Problem of Miracles in the 1830's." *Victorian Studies* 4 (1960):5–32.

———. "The Uniformitarian-Catastrophist Debate." *Isis* 51 (1960):38–55.

———. "The Bases of Darwin's Achievement: A Reevaluation." *Victorian Studies* 5 (1961):109–34.

———. "The Impact of Uniformitarianism: Two Letters from John Herschel to Charles Lyell, 1836–1837." *Proceedings of the American Philosophical Society* 105 (1961):301–14.

———. "The Normative Role of Science in Early Victorian Thought." *Journal of the History of Ideas* 25 (1964):487–502.

———. "History in Depth—The Early Victorian Period." *History of Science* 3 (1964):20–38.

———. "Charles Lyell, Radical Actualism, and Theory." *British Journal for the History of Science* 9 (1976):377–84.

———. Review of *The Comparative Reception of Darwinism*, edited by Thomas F. Glick. *American Historical Review* 8 (1976):559–61.

———. "The Whewell-Darwin Controversy." *Journal of the Geological Society of London* 132 (1976):377–84.

———. *Science in Culture: The Early Victorian Period.* New York: Dawson and Science History Publications, 1978.

Coleman, William. "Lyell and the 'Reality' of Species: 1830–1833." *Isis* 53 (1962):325–28.

———. *Biology in the Nineteenth Century: Problems of Form, Function, and Transformation.* New York: John Wiley, 1971.

Colp, Ralph, Jr. *To Be an Invalid: The Illness of Charles Darwin.* Chicago: University of Chicago Press, 1977.

Cowles, Thomas. "Malthus, Darwin, Bagehot: A Study in the Transference of a Concept." *Isis* 26 (1936):341–48.

Crombie, A. C. "Darwin's Scientific Method." *Actes de IXe Congrès international d'Histoire des Sciences* (1960):354–62.

Darlington, C. D. *Darwin's Place in History.* Oxford: Basil Blackwell, 1959.

———. "The Origin of Darwinism." *Scientific American* 200 (1959):60–66.

Darlington, P. "Darwin and Zoogeography." *Proceedings of the American Philosophical Society* 103 (1959):307–19.

Desmond, R. G. C. "The Hookers and the Development of the Royal Botanic Gardens, Kew." *Biological Journal of the Linnean Society* 7 (1975):173–82.

Dobzhansky, Theodore. "Blyth, Darwin, and Natural Selection." *American Naturalist* 93 (1959):204–6.

Dupree, A. Hunter. *Asa Gray, 1810–1888.* Cambridge, Mass.: Harvard University Press, 1959.

Egerton, Frank N. "Studies of Animal Populations from Lamarck to Darwin." *Journal of the History of Biology* 1 (1968):225–59.

————. "Refutation and Conjecture: Darwin's Response to Sedgwick's Attack on Chambers." *Studies in the History and Philosophy of Science* 1 (1970):176–83.

————. "Humboldt, Darwin and Population." *Journal of the History of Biology* 3 (1970):325–60.

————. "Darwin's Method or Methods?" *Studies in the History and Philosophy of Science* 2 (1971):281–86.

————. "Darwin's Early Reading of Lamarck." *Isis* 67 (1976):452–56.

Eiseley, Loren. *Darwin's Century: Evolution and the Men Who Discovered It.* Garden City, New York: Doubleday, 1958.

————. "Charles Darwin, Edward Blyth, and the Theory of Natural Selection." *Proceedings of the American Philosophical Society* 103 (1959):94–158.

————. *Darwin and the Mysterious Mr. X: New Light on the Evolutionists.* New York: E. P. Dutton, 1979.

Ellegård, Alvar. *Darwin and the General Reader, The Reception of Darwin's Theory of Evolution in the British Periodical Press, 1859–1872.* Goteburg: Almquist and Wiksell, 1958.

Eng, Erling. "Thomas Henry Huxley's Understanding of 'Evolution.' " *History of Science* 16 (1978):291–303.

Farber, Paul. "Buffon and the Concept of Species." *Journal of the History of Biology* 5 (1972):259–84.

Feibleman, James K. "Darwin and Scientific Method." *Tulane Studies in Philosophy* 8 (1959):3–14.

Fleming, Donald. "Charles Darwin, The Anaesthetic Man." *Victorian Studies* 5 (1961):219–36.

Flew, A. G. N. "The Structure of Darwinism." *New Biology* 28 (1959):18–34.

Foster, W. D. "A Contribution to the Problem of Darwin's Ill-health." *Bulletin of the History of Medicine* 39 (1965):476–78.

Freeman, R. B. "Darwin's Negro Bird-Stuffer." *Notes and Records of the Royal Society of London* 33 (1978):83–86.

Freeman, R. B. and Gautrey, P. J. "Darwin's *Questions about the Breeding of Animals,* with a Note on *Queries about Expression.*" *Journal of the Society for the Bibliography of Natural History* 5 (1969):220–25.

————. "Charles Darwin's *Queries about Expression.*" *Bulletin of the British Museum (Natural History)* 4 (1972):207–19.

Fruchtbaum, Harold. *Times Literary Supplement,* October 5, 1967, p. 938. a Letter to the editor.

Gale, Barry G. "Darwin and the Concept of a Struggle for Existence: A Study in the Extrascientific Origins of Scientific Ideas." *Isis* 63 (1972):321–44.

Garfinkel, N. "Science and Religion in England 1790–1800. The Critical Response to the Works of Erasmus Darwin." *Journal of the History of Ideas* 16 (1955):376–88.

Geison, G. L. "Darwin and Heredity: The Evolution of His Hypothesis of Pangenesis." *Journal of the History of Medicine* 24 (1969):375–411.

Ghiselin, Michael T. *The Triumph of the Darwinian Method.* Berkeley and Los Angeles: University of California Press, 1969.

———. "Mr. Darwin's Critics, Old and New." *Journal of the History of Biology* 6 (1973):155–65.

———. "Darwin and Evolutionary Psychology." *Science* 179 (1973):964–68.

———. "Two Darwins: History versus Criticism." *Journal of this History of Biology* 9 (1976):121–32.

Gillespie, Neal. *Charles Darwin and the Problem of Creation.* Chicago: University of Chicago Press, 1979.

Gillispie, C. *Genesis and Geology: The Impact of Scientific Discoveries Upon Religious Beliefs in the Decades Before Darwin.* Cambridge, Mass.: Harvard University Press, 1951.

———. "Lamarck and Darwin in the History of Science." *American Scientist* 46 (1958):388–409.

———. *The Edge of Objectivity: An Essay in the History of Scientific Ideas.* Princeton: Princeton University Press, 1960.

Glass, Bentley, Temkin, Owsei, and Strauss, William L., Jr., eds. *Forerunners of Darwin: 1745–1859.* Baltimore: The Johns Hopkins Press, 1959.

Glick, Thomas F., ed. *The Comparative Reception of Darwin.* Austin: Texas University Press, 1974.

Good, Dr. Rankine. "The Life of the Shawl." *Lancet,* January 9, 1954, pp. 106–7.

———. "The Origin of 'the Origin': A Psychological Approach." *Biology and Human Affairs,* October 1954, pp. 10–16.

Greenacre, Phyllis. *The Quest for the Father: A Study of the Darwin-Butler Controversy, as a Contribution to the Understanding of the Creative Individual.* New York: International Universities Press, Inc., 1963.

Greene, John C. *The Death of Adam.* New York: Mentor Books, 1962.

———. "Reflections on the Progress of Darwin Studies." *Journal of the History of Biology* 8 (1975):243–73.

———. "Darwin as a Social Evolutionist." *Journal of the History of Biology* 10 (1977):1–27.

Grinnell, George. "The Rise and Fall of Darwin's First Theory of Transmutation." *Journal of the History of Biology* 7 (1974):259–73.

Gruber, Howard E., and Gruber, Valmai. "The Eye of Reason: Darwin's Development during the *Beagle* Voyage." *Isis* 53 (1962):186–200.

Gruber, Howard E., and Barrett, Paul H. *Darwin on Man: A Psychological Study of Scientific Creativity.* New York: E. P. Dutton, 1974.

Gruber, Jacob. "Who was the *Beagle*'s Naturalist?" *British Journal for the History of Science* 4 (1969):266–82.

Gunther, A. E. "J. E. Gray, Charles Darwin, and the *Cirripedes*, 1846–

1851." *Notes and Records of the Royal Society of London* 34
(1979):53–63.

Harrison, James. "Erasmus Darwin's View of Evolution." *Journal of the
History of Ideas* 32 (1971):247–64.

Herbert, Sandra. "Research Note: Darwin, Malthus, and Selection."
Journal of the History of Biology 4 (1971):209–17.

———. "The Place of Man in the Development of Darwin's Theory of
Transmutation." 2 parts. *Journal of the History of Biology* 7 (1974):217–
58 and 10 (1977):155–227.

Hermann, Imre. "Charles Darwin." *Imago* 13 (1927):57–82.

Himmelfarb, Gertrude. *Darwin and the Darwinian Revolution.* New
York: W. W. Norton, 1959.

Hodge, M. J. S. "The Universal Gestation of Nature: Chambers' *Vestiges* and *Explanations.*" *Journal of the History of Biology* 5 (1972):127–
51.

———. "The Structure and Strategy of Darwin's 'Long Argument.' "
British Journal for the History of Science 10 (1977):237–46.

Hooykaas, R. "The Principle of Uniformity in Geology, Biology, and
Theology." *Journal of the Transactions of the Victorian Institute*
88 (1956):101–6.

———. *Natural Law and Divine Miracle. A Historical-Critical Study
on the Principle of Uniformity in Geology.* Leiden: Brill, 1963.

Hubble, Dr. Douglas. "Charles Darwin and Psychotherapy." *Lancet,*
January 30, 1943, pp. 129–33.

———. "The Life of the Shawl." *Lancet,* December 26, 1953, pp. 1351–
54.

———. "The Autobiography of Charles Darwin." *Lancet,* July 5, 1958,
pp. 37–39.

———. "Darwin's Illness." *New Statesman,* April 10, 1964, p. 561.

Hull, David. *Darwin and his Critics: The Reception of Darwin's Theory of Evolution by the Scientific Community.* Cambridge, Mass.:
Harvard University Press, 1973.

Huxley, Julian. *Evolution: The Modern Synthesis.* 3rd ed. London: Allen
and Unwin, 1974.

Irvine, William. *Apes, Angels, and Victorians: The Story of Darwin,
Huxley, and Evolution.* New York: McGraw-Hill, 1955.

Jespersen, P. Helveg. "Charles Darwin and Dr. Grant." *Lychnos 1948–
1949* 149 (1950):159–67.

Kass, Leon R. "Teleology and Darwin's *The Origin of Species:* Beyond
Chance and Necessity?" In *Organism, Medicine and Metaphysics:
Essays in Honor of Hans Jonas,* ed. Stuart F. Spicker. Boston: D.
Reidel, 1978.

Kelly, Michael. "Darwin Really Was Sick." *Journal of Chronic Diseases*
20 (1967):341–47.

Kempf, Edward J. "Charles Darwin—the Affective Sources of His Inspiration and Anxiety Neurosis." *Psychoanalytic Review* 5 (1918):151–
92.

———. *Psychopathology.* St. Louis: C. V. Mosby, 1920.

Keynes, R. D. "Darwin and the *Beagle.*" *Proceedings of the American Philosophical Society* 123 (1979):324–35.

King-Hele, D. *Erasmus Darwin.* London: Macmillan, 1963.

Kohn, David. "Theories to Work By: Rejected Theories, Reproduction, and Darwin's Path to Natural Selection." *Studies in the History of Biology* 4 (1980):67–170.

Kohn, Lawrence A. "Charles Darwin's Chronic Ill Health." *Bulletin of the History of Medicine* 37 (1963):239–56.

Krause, Ernst Ludwig. *Erasmus Darwin.* London: John Murray, 1879.

Lankester, Sir E. Ray. "A Great Naturalist—Sir Joseph Hooker." *Annual Report, Smithsonian Institution* (1918):585–601.

Lehman, H. "On the Form of Explanation in Evolutionary Theory." *Theoria* 32 (1966):14–24.

Levin, Samuel. "Malthus and the Idea of Progress." *Journal of the History of Ideas* 27 (1966):92–108.

Levine, George, and Madden, William, eds. *The Art of Victorian Prose.* New York: Oxford University Press, 1968.

Lilley, Samuel. "The Origin and Fate of Erasmus Darwin's Theory of Organic Evolution." *Actes du IXe Congrès international d'Histoire des Sciences* 5 (1965):70–75.

Limoges, Camille. *La selection naturelle: étude sur la première constitution d'un concept (1837–1859).* Paris: Presses Universitaires de France, 1970.

Loewenberg, Bert James. "The Mosaic of Darwinian Thought." *Victorian Studies* 3 (1959):3–18.

———. "Darwin and Darwin Studies, 1959–1963." *History of Science* 4 (1965):15–54.

Lovejoy, Arthur O. "Some Eighteenth Century Evolutionists II." *Popular Science Monthly* 65 (1904):323–40.

———. *The Great Chain of Being: A Study of the History of an Idea.* Cambridge, Mass.: Harvard University Press, 1936.

MacLeod, R. M. "Evolutionism and Richard Owen, 1830–1868: An Episode in Darwin's Century." *Isis* 56 (1965):259–80.

Mandelbaum, M. "Darwin's Religious Views." *Journal of the History of Ideas* 19 (1958):363–78.

Manier, Edward. *The Young Darwin and His Cultural Circle.* Boston: D. Reidel, 1978.

Manser, A. R. "The Concept of Evolution." *Philosophy* 40 (1965):18–34.

Mayr, Ernst. "Open Problems of Darwin Research." *Studies in the History and Philosophy of Science* 2 (1971):273–80.

———. "Lamarck Revisited." *Journal of the History of Biology* 5 (1972):55–94.

———. "Darwin and Natural Selection." *American Scientist* 65 (1977):321–27.

McKinney, H. Lewis. "Alfred Russel Wallace and the Discovery of Natural Selection." *Journal of the History of Medicine and Allied Sciences* 21 (1966):353–57.

———. "Wallace's Earliest Observations on Evolution: 28 December 1845." *Isis* 60 (1969):370–73.

_____. *Wallace and Natural Selection.* New Haven and London: Yale University Press, 1972.

Medawar, D. B. "Darwin's Illness." *Annals of Internal Medicine* 61 (1964):782–86.

Millhauser, Milton. *Just Before Darwin: Robert Chambers and* Vestiges. Middletown, Connecticut: Wesleyan University Press, 1959.

Morrell, J. B. "London Institutions and Lyell's Career: 1820–1841." *British Journal for the History of Science* 9 (1976):132–46.

Mudford, P. G. "Lawrence's Natural History of Man (1819)." *Journal of the History of Ideas* 29 (1968):430–36.

Pantin, C. F. A. "Alfred Russel Wallace, F.R.S., and His Essays of 1858 and 1855." *Notes and Records of the Royal Society of London* 14 (1959):67–84.

Pickering, George. *Creative Malady: Illness in the Lives and Minds of Charles Darwin, Florence Nightingale, Mary Baker Eddy, Sigmund Freud, Marcel Proust, Elizabeth Barrett Browning.* London: Allen, Unwin, 1974.

Review of Auguste Comte's *Cours de Philosophie Positive. Edinburgh Review* 67 (1838):271–308.

Richardson, R. Alan. "Biogeography and the Genesis of Darwin's Ideas on Transmutation." *Journal of the History of Biology* 14 (1981): 1–41.

Roberts, Hyman J. "Reflections on Darwin's Illness." *Geriatrics,* September 1967, pp. 160–67.

Rudwick, Martin J. S. "The Foundation of the Geological Society of London: Its Scheme for Co-operative Research and Its Struggle for Independence." *British Journal for the History of Science* 1 (1963):325–55.

_____. "The Strategy of Lyell's *Principles of Geology.*" *Isis* 61 (1970):5–33.

_____. "Darwin and Glen Roy: A 'Great Failure' in Scientific Method?" *Studies in the History of Philosophy and Science* 5 (1974): 99–185.

_____. "Charles Lyell, F.R.S. (1797–1875) and His London Lectures on Geology 1832–1833." *Notes and Records of the Royal Society of London* 29 (1975):231–63.

_____. "Historical Analogies in the Geological Work of Charles Lyell." *Janus* 64 (1977):89–107.

Ruse, Michael. "The Darwin Industry—A Critical Evaluation." *History of Science* 12 (1974):43–58.

_____. "Charles Darwin's Theory of Evolution: An Analysis." *Journal of the History of Biology* 8 (1975):219–41.

_____. "Natural Selection in the *Origin of Species.*" *Studies in the History and Philosophy of Science* 11 (1971):311–51.

_____. "Charles Darwin and Artificial Selection." *Journal of the History of Ideas* 36 (1975):339–50.

_____. *The Darwinian Revolution.* Chicago: University of Chicago Press, 1979.

Russell-Gebbett, Jean. *Henslow of Hitcham.* Lavenham, Suffolk: Terence Dutton, 1977.

Schneer, Cecil J., ed. *Toward a History of Geology.* Cambridge, Mass.: M.I.T. Press, 1969.

Schwartz, Joel S. "Charles Darwin's Debt to Malthus and Edward Blyth."
    Journal of the History of Biology 7 (1974):301–18.
Schweber, Silvan S. "The Origin of the *Origin* Revisited." *Journal of
    the History of Biology* 10 (1977):223–316.
——. "Early Victorian Science: *Science in Culture.*" *Journal of the
    History of Biology* 13 (1980):121–40.
——. "Darwin and the Political Economists: Divergence of Charac-
    ter." *Journal of the History of Biology* 13 (1980):195–289.
Scott, Robert H. "The History of Kew Observatory." *Proceedings of the
    Royal Society* 39 (1885):37–86.
Secord, James A. "Nature's Fancy: Charles Darwin and the Breeding of
    Pigeons." *Isis* 72 (1981):163–86.
Simpson, G. G. *The Meaning of Evolution.* New Haven: Yale Univer-
    sity Press, 1949.
Smith, C. U. M. "Charles Darwin, the Origin of Consciousness, and
    Panpsychism." *Journal of the History of Biology* 11 (1978):245–67.
Smith, Sydney. "The Origin of the *Origin,* As Discerned from Charles
    Darwin's Notebooks and His Annotations in the Books He Read
    Between 1837 and 1842." *The Advancement of Science* 64 (1960):391–
    402.
——. "The Darwin Collection at Cambridge with One Example of Its
    Use: Charles Darwin and *Cirripedes.*" *Acte du IXe Congrès inter-
    national d'Histoire des Sciences* (1965):96–100.
Stecher, Robert M. "Darwin-Innes Letters: The Correspondence of an
    Evolutionist with His Vicar, 1848–1884." *Annals of Science* 17
    (1961):201–58.
Stoddart, D. R. "Darwin, Lyell, and the Geological Significance of Coral
    Reefs." *British Journal for the History of Science* 9 (1976): 199–218.
Sulloway, Frank J. "Geographic Isolation in Darwin's Thinking: The
    Vicissitudes of a Crucial Idea." *Studies in the History of Biology* 3
    (1979):23–65.
Turril, William B. *Pioneer Plant Geography: The Phytogeographical Re-
    searches of Sir Joseph Hooker.* The Hague: M. Nijhoff, 1953.
Vorzimmer, Peter. "Charles Darwin and Blending Inheritance." *Isis* 54
    (1963):371–90.
——. "Darwin, Malthus, and the Theory of Natural Selection." *Jour-
    nal of the History of Ideas* 30 (1969):527–42.
——. "Darwin's *Questions About the Breeding of Animals* (1839)."
    Journal of the History of Biology 2 (1969):269–81.
——. *Charles Darwin: The Years of Controversy: The Origin of Spe-
    cies and Its Critics, 1859–1882.* Philadelphia: Temple University
    Press, 1970.
——. "The Darwin Reading Notebooks (1838–1860)." *Journal of the
    History of Biology* 10 (1977):107–53.
Wichler, Gerhard. *Charles Darwin: The Founder of the Theory of Evo-
    lution and Natural Selection.* New York: Pergamon Press, 1961.
Wilson, Leonard G. *Sir Charles Lyell's Scientific Journals on the Spe-
    cies Question.* New Haven: Yale University Press, 1970.

———. "Sir Charles Lyell and the Species Question." *American Scientist* 49 (1971):43–55.

———. *Charles Lyell: The Years to 1841.* New Haven and London: Yale University Press, 1972.

———. "Geology on the Eve of Charles Lyell's First Visit to America, 1841." *Proceedings of the American Philosophical Society* 124 (1980): 168–202.

Winslow, John H. *Darwin's Victorian Malady: Evidence for its Medically Induced Origin.* Philadephia: American Philosophical Society, 1971.

Wood, Roger J. "J. Robert Bakewell (1725–1795), Pioneer Animal Breeder and His Influence on Charles Darwin." *Folia Mendeliana* 8 (1973):231–42.

Woodruff, A. W. "Darwin's Health in Relation to His Voyage to South America." *British Medical Journal* 1 (1965):745–50.

———. "The Impact of Darwin's Voyage to South America on His Work and Health." *Bulletin of the New York Academy of Medicine* 44 (1968):661–72.

Young, Robert. "Malthus and the Evolutionists: The Common Context of Biological and Social Theory." *Past and Present* 43 (1969):109–45.

Index

editor, 34; as reference, 123, 130;
contrasted to Darwin, 159;
contrasted to Hooker, 55; coral
theory of, 43–44; debt to, 44–45,
162; distribution and, 67;
friendship with, 41–48; geology
and, 33–54; geology theory of, 37;
influence by, 25, 26, 53, 146, 160;
interest in Darwin, 39, 41, 115;
letters from, 40, 44, 46, 47; letters
to, 30, 41, 44, 46, 48, 57, 95, 116,
on Forbes, 67, on geology, 45, on
Journal, 51–52, on priority, 95, on
publishing, 96, on Wallace letter,
148; support by, 98; support for,
147; uniformitarianism and,
36–39; use of analogy by, 37; visits
to, 48

McCormick, Robert, 19
MacGillivray, William, 11
McKinney, H. Lewis, 155
Macleay, W. S., 111
Madeira, 67
Maer, 13, 14, 17, 49
magnetic survey work, 56
Malacca, 114
Malthus, 3, 5, 7, 101, 122, 143, 158,
160; as a reference, 137, 165;
disagreement with, 104, 111;
influence by, 3–4, 31, 152; place in
studies of, 7, 30, 31
Malvern, 146
mammalia, 43, 68, 107, 132, 134
mammoth, 23
management, 5, 89, 152; of people,
79, 85–89, 145
marriage, 4, 40, 49
mastodon, 165
materialism, 8, 107, 109, 159, 160
materia medica, 10
mathematics, 13, 79
matter, 107, 109, 114, 136
Mayr, Ernst, 155
"medical evolutionists," 156
medicine, 10, 11, 12, 13, 56
Megatherium, 22
mental activity, 102, 106, 108
metaphor: use of, 101, 123, 130
meteorology, 20, 57
microscope, 12
midwifery, 10
migration, 71, 131, 132, 133, 139,
166, 168

mineralogists, 82
mineralogy, 15, 39
Miocene, 68
mite, the, 107
M Notebook, 108, 109, 122
modification, 72, 126, 133, 134
mollusca, 66, 107
Monro, Alexander, 10
Monte Video, 21
Moor Park, 94
morphology, 31, 114, 133, 134, 139,
169
mountains, 37, 47, 132
multiplying effect, 103, 123, 129, 159
Murchison, Roderick Impey, 130
Muscicapa coronata, 106
Museum of Natural History
(Edinburgh), 11
museums, 130
music: Darwin's interest in, 13
mutability, 111, 135, 137, 139. *See
also* immutability

natural history: accounts of, 155; at
Cambridge, 17; career in, 30, 144;
Darwin on, 90, 150; education in,
9; Hooker and, 74; interest for, 10,
23, 25, 244; knowledge of, 31;
propositions for, 85; skills for, 8,
12
Natural History of Selborne (White), 9
naturalists, 9, 34, 77, 82, 99, 135,
137; as surgeons, 56; *Beagle*
position of, 17–18, 19, 144;
ignorance by, 120, 122, 123, 165;
on species, 74; questions for, 107;
utilitarianism and, 128
natural laws, 105, 107. *See also* laws
natural philosophy, 158
Natural Sciences, 15, 52, 96
natural selection, 71, 93, 94, 97, 113,
126, 131, 135, 137, 149; as
evidence, 166, 167; Creationists
and, 122, 124–25; evidence for,
133, 138, 169; goal of, 136;
influence of, 3; instinct in, 127,
128; in the *Origin*, 120, 123, 124;
migration and, 133; perfection and,
127. *See also* selection
natural theology, 159
"*natura non facit saltum*," 129, 167
Nature, 79
nature: antagonism/cooperation in,
158; certainties of, 165